MONOGRAPHS ON
APPLIED PROBABILITY AND STATISTICS

FINITE MIXTURE DISTRIBUTIONS

MONOGRAPHS ON
APPLIED PROBABILITY AND STATISTICS

General Editor

D.R. COX, FRS

Also available in the series

Probability, Statistics and Time
M.S. Bartlett

The Statistical Analysis of Spatial Pattern
M.S. Bartlett

Stochastic Population Models in Ecology and Epidemiology
M.S. Bartlett

Risk Theory
R.E. Beard, T. Pentikäinen and E. Pesonen

Point Processes
D.R. Cox and V. Isham

Analysis of Binary Data
D.R. Cox

The Statistical Analysis of Series of Events
D.R. Cox and P.A.W. Lewis

Queues
D.R. Cox and W.L. Smith

Stochastic Abundance Models
E. Engen

The Analysis of Contingency Tables
B.S. Everitt

Finite Mixture Distributions
B.S. Everitt and D.J. Hand

Population Genetics
W.J. Ewens

Classification
A.D. Gordon

Monte Carlo Methods
J.M. Hammersley and D.C. Handscomb

Identification of Outliers
D.M. Hawkins

Multivariate Analysis in Behavioural Research
A.E. Maxwell

Applications of Queueing Theory
G.F. Newell

Some Basic Theory for Statistical Inference
E.J.G. Pitman

Statistical Inference
S.D. Silvey

Models in Regression and Related Topics
P. Sprent

Sequential Methods in Statistics
G.B. Wetherill

Finite Mixture Distributions

B.S. EVERITT
Head of Biometrics Unit, Institute of Psychiatry
and
D.J. HAND
Lecturer, Biometrics Unit, Institute of Psychiatry

LONDON NEW YORK
CHAPMAN AND HALL

First published in 1981 by
Chapman and Hall Ltd
11 New Fetter Lane, London EC4P 4EE

Published in the USA by
Chapman and Hall
in association with Methuen, Inc.
733 Third Avenue, New York, NY 10017

Printed in Great Britain
at the University Press, Cambridge

ISBN 0 412 22420 8

British Library Cataloguing in Publication Data

Everitt, B S
Finite mixture distributions. – (Monographs
on applied probability and statistics).

1. Distribution (Probability theory)
I. Title II. Series
519.5′32 QA273.6 80–41131

ISBN 0–412–22420–8

Contents

Preface

Finite mixture distributions arise in a variety of applications ranging from the length distribution of fish to the content of DNA in the nuclei of liver cells. The literature surrounding them is large and goes back to the end of the last century when Karl Pearson published his well-known paper on estimating the five parameters in a mixture of two normal distributions. In this text we attempt to review this literature and in addition indicate the practical details of fitting such distributions to sample data. Our hope is that the monograph will be useful to statisticians interested in mixture distributions and to research workers in other areas applying such distributions to their data. We would like to express our gratitude to Mrs Bertha Lakey for typing the manuscript.

Institute of Psychiatry
University of London
1980

B.S. Everitt
D.J. Hand

CHAPTER 1

General introduction

1.1 Introduction

This monograph is concerned with statistical distributions which can be expressed as superpositions of (usually simpler) component distributions. Such superpositions are termed *mixture distributions* or *compound distributions*. For example, the distribution of height in a population of children might be expressed as follows:

$$h(\textit{height}) = \int g(\textit{height}\,;\,\textit{age})\,f(\textit{age})\,\mathrm{d}\,\textit{age} \qquad (1.1)$$

where $g(\textit{height}; \textit{age})$ is the conditional distribution of height on age, and $f(\textit{age})$ is the age distribution of the children in the population. The probability density function of height has been expressed as an infinite superposition of conditional height density functions, and is thus a mixture density. As a further example, we might express the height density function in the form

$$h(\textit{height}) = h_1(\textit{height}\,;\,\textit{boy})\,p(\textit{boy}) + h_2(\textit{height}\,;\,\textit{girl})\,p(\textit{girl}) \qquad (1.2)$$

where $p(\textit{boy})$ and $p(\textit{girl})$ are, respectively, the probabilities that a member of the population is a boy or a girl, and h_1 and h_2 are the height density functions for boys and girls. Thus the density function of height has been expressed as a superposition of two conditional density functions.

Density functions of the forms (1.1) and (1.2) have received increasing attention in the statistical literature recently, partly because of interest in their mathematical properties, but mainly because of the considerable number of areas in which such density functions are encountered. In this text we are primarily concerned with mixtures such as (1.2), involving a finite number of components, and which, not surprisingly, are termed finite mixtures. The problems of central interest arise when data are not available for each conditional distribution separately, but only for the overall mixture distribution. Often such situations arise because it is impossible to observe some underlying variable which splits the observations into groups – only the combined distribution can be studied. In these circumstances interest often focuses on estimating the *mixing proportions* [$p(boy)$ and $p(girl)$ in (1.2)] and on estimating the parameters in the conditional distributions. We concentrate most of our effort in these areas but also consider other topics of practical importance, such as ways of determining when a data set arises from a mixture and when it does not. However, to illustrate the wide range of areas in which mixture distributions occur we first examine some examples.

1.2 Some applications of finite mixture distributions

The most widely used finite mixture distributions are those involving normal components. Medgyessi (1961) analyses absorption spectra in terms of normal mixtures – 'to every theoretical "line" belongs an intensity distribution whose graph fits very well to that of some normal distribution' – and also applies normal mixtures to the results of protein separation by electrophoresis. Bhattacharya (1967) studies the length distribution of a certain type of fish and finds it useful to split his observations into age categories, with each category contributing a normal component distribution to yield an overall mixture. Cassie (1954) has studied the same problem. Gregor (1969) applies a mixture of normal distributions to data arising from measuring the content of DNA in the nuclei of liver cells of rats. Such a distribution is thought appropriate in this case because in some organs there exist various classes of nuclei of cells which have characteristic differences in DNA content. Holgersson and Jorner (1976) compare various estimation schemes for normal mixtures on fibre size distributions of myelinated lumbosacral ventral root fibres from kittens of various ages. Clark *et al.* (1968) provide an illustration of an area in which mixture distributions are being applied more frequently, namely the study of disease distributions. The general question is: is there more than one type of disease here? Clark *et al.*, studying hypertension, investigate whether a sample of blood pressure data can be separated into two normal populations. Normal mixtures have also been used in many studies designed to investigate the robustness of certain statistical techniques when the data are not normally distributed. Some examples of this are given by Subrahmaniam, Subrahmaniam and Messeri (1975), Hyenius (1950), and Weibull (1950).

Another general area where mixtures of distributions are important is in failure data. Here the observations are the times to failure of a sample of items. Often failure can occur for more than one reason (see, for example, Davis, 1952) and the failure distribution for each reason can be adequately approximated by a simple density function such as the negative exponential. The overall failure distribution is then a mixture. Several attempts have been made to fit such mixtures to the failure distribution of electronic valves. For example, Mendenhall and Hader (1958) fit exponential components and Kao (1959) fits Weibull components.

'Failure' can of course be interpreted in a perfectly general way as 'event' and Thomas (1966) has used an exponential mixture to model the distribution of the time to discharge of nerve cells. Ashton (1971) has used a mixture of a gamma distribution with a displaced exponential distribution to model the frequency distribution of time gaps in road traffic. Joffe (1964) used a mixture with the components of the form

$$g(x) = r\exp(-s\sqrt{x})$$

to fit the size frequency distribution of dust particles in mines.

Discrete mixtures are applied by Medgyessi (1961) to the counter current method of identifying the constituents of small amounts of organic chemicals. This involves sequentially diffusing the dissolved chemical mixture into a number of cells containing fresh solvent. The result of this exercise is that each component of the chemical mixture is distributed independently of the others according to binomial distributions across the cells (for details see Medgyessi, 1961, or Craig, 1944). The final result is a binomial mixture.

Blischke (1978, Vol. 1, p. 175) gives an example of a mixture of two trinomial distributions, namely the sex distribution of twin pairs. A twin pair must be one of three classes: male/male, male/female, female/female. Thus we are dealing with a trinomial distribution. However, twin pairs come in two types, dizygotic and monozygotic, so that we have in fact a mixture of two trinomials. For the dizygotic class the probabilities of the three classes are p^2, $2pq$, q^2, and for the monozygotic class p, 0, q. The mixing proportions are the relative magnitudes of the two classes of twin pairs. Latent class analysis (Lazarsfeld, 1968) may also be viewed from a discrete mixture viewpoint (see Section 4.5).

1.3 Definition

Section 1.1 gave an informal description and Section 1.2 gave illustrative examples of the density functions and distributions with which we are concerned. In this section we begin with a more formal definition of a mixture density and outline a problem which can occur with distributions of this kind.

First, then, a definition:

Definition: Let $g(\boldsymbol{x};\boldsymbol{\theta})$ be a d-dimensional probability density function depending on an m-dimensional parameter vector $\boldsymbol{\theta}$ and let $H(\boldsymbol{\theta})$ be an m-dimensional cumulative distribution function. Then

$$f(\boldsymbol{x}) = \int g(\boldsymbol{x};\boldsymbol{\theta})\,\mathrm{d}H(\boldsymbol{\theta}) \tag{1.3}$$

is called a *mixture density*. H is called the *mixing distribution*. □

From this definition it is clear that $f(\boldsymbol{x})$ can be viewed as being a marginal density of a $(d+m)$-variate density. One consequence of this is that any density function $f(\boldsymbol{x})$ can be viewed as a mixture density simply by imagining extra variables which have been integrated over.

The definition given above is perfectly general but most of the applications are concerned with a subset of this general definition. This subset is generated by the special case when H is discrete and assigns positive probability to only a finite number of points $\{\boldsymbol{\theta}_i;\, i=1,\ldots,c\}$. When this is the case we can replace the integral in (1.3) by a sum to give the *finite mixture*

$$f(\boldsymbol{x}) = \sum_{i=1}^{c} H_i(\boldsymbol{\theta}_i)\,g(\boldsymbol{x};\boldsymbol{\theta}_i). \tag{1.4}$$

This text is concerned with finite mixtures and (1.4) will serve as a canonical form throughout. Matching definitions can be given for mixture *distributions* (see, for example, Yakowitz and Spragins, 1968).

The parameters in expression (1.4) can be divided into three types. The first consists solely of c, the number of components of the finite mixture. The second consists of the mixing proportions $H_i(\boldsymbol{\theta}_i)$ (which we shall usually denote by p_i). The third consists of the component parameter vectors $\boldsymbol{\theta}_i$. General methods for estimating parameters from each of these three classes will be presented below, but first we consider a fundamental difficulty which can complicate parameter estimation in some types of mixture distribution.

Suppose that we had a set of samples known to come from a mixture of two univariate uniform distributions and that our aim was to analyse the mixture into its components. Our analysis might produce the mixture

$$f(x) = \tfrac{1}{3}U(-1, 1) + \tfrac{2}{3}U(-2, 2)$$

(where $U(a, b)$ is a uniform distribution with range (a, b)).

However, given that we only have samples from the mixture, and not from each component separately, there is no way we could tell if this decomposition was correct or if

$$f(x) = \tfrac{1}{2}U(-2, 1) + \tfrac{1}{2}U(-1, 2)$$

was the correct decomposition. The problem is clearly serious if the ultimate aim requires knowledge of the individual components (as, for example, in the case when we intend to classify future observations into one of the two classes from our knowledge of the component distributions) but even in situations where the component structure of the overall density is not important the existence of multiple solutions can make the estimation of parameters much more difficult.

This problem, which first appears to have been studied by Teicher (1961, 1963) is known as that of the non-identifiability of a class of mixtures. Formally, we have

Definition: A class D of mixtures is said to be *identifiable* if and only if for all $f(\boldsymbol{x}) \in \mathrm{D}$ the equality of two representations

$$\sum_{i=1}^{c} p_i g_i(\boldsymbol{x}; \boldsymbol{\theta}_i) = \sum_{j=1}^{c'} \hat{p}_j g_j(\boldsymbol{x}; \hat{\boldsymbol{\theta}}_j)$$

implies that $c = c'$ and for all i there exists some j such that $p_i = \hat{p}_j$ and $\boldsymbol{\theta}_i = \hat{\boldsymbol{\theta}}_j$. □

In practice non-identifiability can be a problem with discrete distributions since if there are m categories we cannot set up more than $(m - 1)$ independent equations and hence can only determine this number of parameters. Consider, for example, a Bernoulli distribution formed from a mixture of two Bernoulli components with probability functions

	$P(x = 0)$	$P(x = 1)$
Component 1	$g_1(0; \theta_1)$	$g_1(1; \theta_1)$

Component 2 $\quad g_2(0;\theta_2) \quad g_2(1;\theta_2)$

Then

$$f(x) = p_1 g_1(x;\theta_1) + p_2 g_2(x;\theta_2).$$

Unfortunately we can only establish a single equation

$$f(0) = p_1 g(0;\theta_1) + p_2 g_2(0;\theta_2)$$

from this since $f(1) = 1 - f(0)$. We have, however, three independent parameters to estimate, namely p_1, $g(0;\theta_1)$, and $g(0;\theta_2)$. [Note that $p_2 = 1 - p_1$, $g(1;\theta_1) = 1 - g_1(0;\theta_1)$, $g_2(1;\theta_2) = 1 - g_2(0;\theta_2)$.]

Fortunately, for continuous distributions the problem seldom arises.

Yakowitz and Spragins (1968) present a useful theorem which helps to show which distributions yield identifiable finite mixtures.

Theorem: A necessary and sufficient condition that the class of all finite mixtures of the set

$$\{G(\boldsymbol{x};\boldsymbol{\theta});\boldsymbol{x}\in\mathscr{R}^d,\ \boldsymbol{\theta}\in\mathscr{R}^m\}$$

be identifiable is that this set should be linearly independent over the real numbers $\mathscr{R}$.

Proof

Necessity

Begin by assuming that the $G(\boldsymbol{x};\boldsymbol{\theta}_i)$ are *not* linearly independent so that there exists some linear function

$$\sum_{i=1}^{c} p_i G(\boldsymbol{x};\boldsymbol{\theta}_i) = 0$$

with $G(\boldsymbol{x};\boldsymbol{\theta}_i) \neq G(\boldsymbol{x};\boldsymbol{\theta}_j)$ if $i \neq j$. Without loss of generality we can assume that the p_i are ordered so that $p_i < 0$ if and only if $i \leqslant c'$.

Then

$$\sum_{i=1}^{c'} |p_i| G(\boldsymbol{x};\boldsymbol{\theta}_i) = \sum_{i=c'+1}^{c} |p_i| G(\boldsymbol{x};\boldsymbol{\theta}_i).$$

Since the $G(\boldsymbol{x};\boldsymbol{\theta}_i)$ are cumulative distribution functions (c.d.f.s) we have that $G(\infty;\boldsymbol{\theta}_i) = 1$ and hence

$$\sum_{i=1}^{c'} |p_i| = \sum_{i=c'+1}^{c} |p_i| = b > 0.$$

Defining $a_i = |p_i|/b$ yields

$$\sum_{i=1}^{c'} a_i G(\boldsymbol{x};\boldsymbol{\theta}_i) = \sum_{i=c'+1}^{c} a_i G(\boldsymbol{x};\boldsymbol{\theta}_i)$$

both being finite mixtures and yet being distinct. That is, they are not identifiable.

Sufficiency

If the class of finite mixtures defined above was non-identifiable then there would exist some mixture such that

$$\sum_{i=1}^{c} p_i G(\boldsymbol{x};\boldsymbol{\theta}_i) = \sum_{j=1}^{c'} p_j G(\boldsymbol{x};\boldsymbol{\theta}_j) \quad (1.5)$$

with not all p_i equal to some p_j and/or not all $\boldsymbol{\theta}_i$ equal to some $\boldsymbol{\theta}_j$. Expression (1.5) could then be rewritten as

$$\sum_{i=1}^{c''} p_i G(\boldsymbol{x};\boldsymbol{\theta}_i) = 0$$

with *no* $G(\boldsymbol{x};\boldsymbol{\theta}_i)$ equal to any $G(\boldsymbol{x};\boldsymbol{\theta}_j)$ and no $p_i = 0$. That is, the $G(\boldsymbol{x};\boldsymbol{\theta}_i)$ are not linearly independent. □

Teicher (1961, 1963) and Yakowitz and Spragins (1968) have used this and other theorems to demonstrate the identifiability of finite mixtures of multidimensional normal c.d.f.s, exponential c.d.f.s, Cauchy c.d.f.s, and others.

Chandra (1977) has investigated the problem of identifiability in more depth and presents some very interesting theorems.

1.4 Estimation methods

Many methods have been devised and used for estimating the parameters of mixture distributions, ranging from Pearson's (1894) method of moments, through formal maximum likelihood approaches, to informal graphical techniques. In this section we outline some methods which are of general applicability. Methods which are restricted to particular types of components, application of the general methods to particular components, and methods which have been developed for other special cases are dealt with in subsequent chapters.

1.4.1 *Maximum likelihood*

In this section we introduce the maximum likelihood method for estimating the parameters of statistical distributions.

Not only is it appealing on intuitive grounds, but it also possesses desirable statistical properties. For example, under very general conditions the estimators obtained by the method are consistent (they converge in probability to the true parameter values) and they are asymptotically normally distributed.

We begin by assuming that $\boldsymbol{x}_1, \ldots, \boldsymbol{x}_n$ are independent observations from a density $f(\boldsymbol{x};\boldsymbol{\alpha})$ where $\boldsymbol{\alpha}$ is the parameter vector we wish to estimate. The joint probability density of $\{\boldsymbol{x}_1, \ldots, \boldsymbol{x}_n\} = \boldsymbol{X}^n$ is

$$\mathscr{L}(\boldsymbol{X}^n;\boldsymbol{\alpha}) = \prod_{i=1}^{n} f(\boldsymbol{x}_i;\boldsymbol{\alpha}).$$

Now, instead of regarding $\mathscr{L}$ as a function of the $\boldsymbol{x}_i$ we can regard it as a function of $\boldsymbol{\alpha}$. It is then called the *likelihood function* – it measures the relative likelihood that different $\boldsymbol{\alpha}$ will have given rise to the observed $\boldsymbol{X}^n$. We can thus try to find that particular $\boldsymbol{\alpha}(=\boldsymbol{\alpha}_0)$ which maximizes $\mathscr{L}$, i.e. that $\boldsymbol{\alpha}_0$ such that the observed $\boldsymbol{X}^n$ is more likely to have come from $f(\boldsymbol{x};\boldsymbol{\alpha}_0)$ than $f(\boldsymbol{x};\boldsymbol{\alpha})$ for any other value of $\boldsymbol{\alpha}$.

For many parameter estimation problems one can tackle this maximization in the traditional way of differentiating $\mathscr{L}$ with respect to the components of $\boldsymbol{\alpha}$ and equating the derivatives to zero to give the *normal equations*

$$\frac{\partial \mathscr{L}}{\partial \alpha_i} = 0.$$

These are then solved for the α_i and the second-order derivatives are examined to verify that it is indeed a maximum which has been achieved and not some other stationary point.

For mixture distributions, however, things are slightly more complicated, and a little thought has to be given as to whether this is the right way to proceed. First, for mixture distributions the normal equations are not usually explicitly solvable for the α_i and so iterative (hill-climbing) techniques have to be adopted. One might consider abandoning the normal equations altogether and simply using hill-climbing techniques to maximize $\mathscr{L}$. The consequences of this are discussed immediately below.

The second extra difficulty raised by the distribution being a mixture is that for mixtures the likelihood function is often unbounded, as is $\partial\mathscr{L}/\partial\alpha_i$. This therefore raises the question of whether the local maximum given by the normal equations (assuming it is a maximum and not some other stationary value) is the one we want to find, or whether (one of) the $\mathscr{L}(\boldsymbol{X}^n;\boldsymbol{\alpha})=\infty$ solutions would be preferable (and these would presumably be found by using hill-climbing to maximize $\mathscr{L}$ as suggested above).

Some insight into this question can be gained by examining the $\boldsymbol{\alpha}$ values for which $\mathscr{L}(\boldsymbol{X}^n;\boldsymbol{\alpha})=\infty$. As an example (which is returned to in Chapter 2), consider a one-dimensional normal mixture. If we take $\mu_c=x_1$ and let $\sigma_c\to 0$ then $\mathscr{L}\to\infty$. It is clear that, in this case at least, the $\mathscr{L}(\boldsymbol{X}^n;\boldsymbol{\alpha})=\infty$ solutions are pathological for our purposes: they do not yield useful mixtures. Note that if the variances in this normal example (and the corresponding spread parameters in other mixtures) were fixed or indeed proportional to each other, the problem would not arise. Kazakos (1977), for example, considers the case when only the p_i are unknown and shows that the likelihood function is concave. Imposing such restrictions is a satisfactory solution in some cases but in general it is obviously not so. When the latter is true the solution usually adopted is to find the largest *stationary* maximum. We are thus led back to the normal equations – using iterative methods to solve them (and *not*, as suggested above, to find the maximum of $\mathscr{L}$).

If, to distinguish clearly between the mixing proportion p_i and the component parameters $\boldsymbol{\theta}_i$, we write $\boldsymbol{\alpha}=(\boldsymbol{p},\boldsymbol{\theta}_1,\boldsymbol{\theta}_2,\ldots,\boldsymbol{\theta}_c)$, we have

$$\mathscr{L}(\boldsymbol{X}^n;\boldsymbol{p},\boldsymbol{\theta}_1,\ldots,\boldsymbol{\theta}_c)=\prod_{j=1}^{n}\left[\sum_{i=1}^{c}p_i g_i(\boldsymbol{x}_j;\boldsymbol{\theta}_i)\right].$$

Rather than differentiating this and equating it directly to zero, we shall define

$$L(\boldsymbol{X}^n;\boldsymbol{p},\boldsymbol{\theta}_1,\ldots,\boldsymbol{\theta}_c)=\log\mathscr{L}(\boldsymbol{X}^n;\boldsymbol{p},\boldsymbol{\theta}_1,\ldots,\boldsymbol{\theta}_c)-\lambda\left(\sum_{i=1}^{c}p_i-1\right).$$

This introduces the constraint $\sum_{i=1}^{c}p_i=1$ via the Lagrange multiplier λ and it also makes use of the fact that since log is a monotonic transformation L will take its maxima at the same parameter values that $\mathscr{L}$ does (the log transformation simply eases subsequent algebra). This gives

$$\frac{\partial L}{\partial p_k}=\sum_{j=1}^{n}\frac{g_k(\boldsymbol{x}_j;\boldsymbol{\theta}_k)}{f(\boldsymbol{x}_j)}-\lambda=0 \tag{1.6}$$

and

$$\frac{\partial L}{\partial \theta_{ik}} = \sum_{j=1}^{n} p_k \frac{\partial g_k(\boldsymbol{x}_j\,;\boldsymbol{\theta}_k)/g\theta_{ik}}{f(\boldsymbol{x}_j)} = 0. \tag{1.7}$$

The Lagrange multiplier λ can be found by multiplying (1.6) by p_k and summing over k to give

$$n - \lambda = 0.$$

Moreover, since by Bayes's theorem the probability that a given $\boldsymbol{x}_j$ comes from component k is

$$\Pr(k \mid \boldsymbol{x}_j) = \frac{p_k g_k(\boldsymbol{x}_j;\boldsymbol{\theta})}{f(\boldsymbol{x}_j)}, \tag{1.8}$$

if we multiply (1.6) by p_k we can express the maximum likelihood estimate $\hat{p}_k$ in the form

$$\hat{p}_k = \frac{1}{n} \sum_{j=1}^{n} \Pr(k \mid \boldsymbol{x}_j); \tag{1.9}$$

i.e. the maximum likelihood estimate of the mixing proportions for class k is given by the sample mean of the conditional probabilities that the $\boldsymbol{x}_j$ come from class k.

Also, using (1.8) in (1.7) we can express (1.7) as

$$\sum_{j=1}^{n} \Pr(k \mid \boldsymbol{x}_j) \frac{\partial \log g_k(\boldsymbol{x}_j\,;\boldsymbol{\theta}_k)}{\partial \theta_{ik}} = 0 \tag{1.10}$$

i.e. the maximum likelihood equations for estimating the parameters $\boldsymbol{\theta}$ are weighted averages of the maximum likelihood equations

$$\partial \log g_k(\boldsymbol{x}_j\,;\hat{\boldsymbol{\theta}}_k)/\partial \theta_{ik} = 0$$

arising from each component considered separately. The weights are the probabilities of membership of the $\boldsymbol{x}_j$ in each class.

Equations (1.9) and (1.10) can provide the basis for an iterative solution of the normal equations (see, for example, Hasselblad, 1969) but Dempster, Laird, and Rubin (1977) (in a paper chiefly concerned with maximum likelihood estimation, although the method can also be applied to Bayesian estimation) show that this is a special case of their EM algorithm. This algorithm has two steps, the first being to estimate the membership probabilities of each observation for each component and the second being equivalent to c separate estimation problems with each observation contributing to the log-likelihood associated with each of the c components with a weight given by the estimated membership probability. These two steps are then repeated iteratively. A further example applied to mixture distributions is given by Wolfe (1969).

Other estimation methods for solving the normal equations include the classical ones of scoring for parameters and the Newton–Raphson method (see Chapter 2). Subsequent chapters will give details of the methods.

1.4.2 *Bayesian estimation*

The general Bayesian approach estimates the *a posteriori* probability density function (p.d.f.) of the parameter vector $\boldsymbol{\alpha}$ through Bayes theorem:

$$f(\boldsymbol{\alpha}|\boldsymbol{X}^n) = \frac{\mathscr{L}(\boldsymbol{X}^n;\boldsymbol{\alpha})f(\boldsymbol{\alpha})}{\displaystyle\int \mathscr{L}(\boldsymbol{X}^n;\boldsymbol{\alpha})f(\boldsymbol{\alpha})\mathrm{d}\boldsymbol{\alpha}}. \tag{1.11}$$

Note that $f(\boldsymbol{x}), f(\boldsymbol{x}_n;\boldsymbol{\alpha}), f(\boldsymbol{\alpha})$, and $f(\boldsymbol{\alpha}|\boldsymbol{X}^n)$, etc. are different functions but confusion will not arise because the arguments will always be explicitly stated. This multiple use of the same symbol greatly simplifies the notation.

Expression (1.11) can be rewritten into the sequential form

$$f(\boldsymbol{\alpha}|\boldsymbol{X}^n) = \frac{f(\boldsymbol{x}_n;\boldsymbol{\alpha})f(\boldsymbol{\alpha}|\boldsymbol{X}^{n-1})}{\displaystyle\int f(\boldsymbol{x}_n;\boldsymbol{\alpha})f(\boldsymbol{\alpha}|\boldsymbol{X}^{n-1})\mathrm{d}\boldsymbol{\alpha}}. \tag{1.12}$$

In attempting to apply this approach to estimating mixture distributions two major difficulties are encountered. The first is the problem of identifiability discussed above. In the present context we can have

$$f(\boldsymbol{x}|\boldsymbol{X}^n) \rightarrow f(\boldsymbol{x}) \text{ while } g_i(\boldsymbol{x}|\boldsymbol{X}^n) \nrightarrow g_i(\boldsymbol{x};\boldsymbol{\theta}_i).$$

As we have already commented, whether or not this is a serious problem will depend on the application.

The second difficulty, perhaps the more important one, is that of the feasibility of the computations involved. In general, when applying Bayes theorem to parameter estimation considerable simplifications can result if the statistic $\hat{\boldsymbol{\alpha}}$ used to estimate $\boldsymbol{\alpha}$ is *sufficient*. A sufficient statistic is one for which $\mathscr{L}(\boldsymbol{x}^n;\hat{\boldsymbol{\alpha}},\boldsymbol{\alpha})$ is independent of $\boldsymbol{\alpha}$. That is, a sufficient statistic satisfies

$$\mathscr{L}(\boldsymbol{X}^n;\boldsymbol{\alpha}) = f(\hat{\boldsymbol{\alpha}};\boldsymbol{\alpha})\mathscr{L}(\boldsymbol{X}^n;\hat{\boldsymbol{\alpha}}) \tag{1.13}$$

in place of the general

$$\mathscr{L}(\boldsymbol{X}^n;\boldsymbol{\alpha})=f(\hat{\boldsymbol{\alpha}};\boldsymbol{\alpha})\mathscr{L}(\boldsymbol{X}^n;\hat{\boldsymbol{\alpha}},\boldsymbol{\alpha}).$$

The importance of this is seen when we substitute (1.13) into (1.11):

$$f(\boldsymbol{\alpha}|\boldsymbol{X}^n)=\frac{f(\hat{\boldsymbol{\alpha}},\boldsymbol{\alpha})f(\boldsymbol{\alpha})}{\int f(\hat{\boldsymbol{\alpha}},\boldsymbol{\alpha})f(\boldsymbol{\alpha})\,\mathrm{d}\boldsymbol{\alpha}}$$

which can lead to considerable simplification if $f(\hat{\boldsymbol{\alpha}},\boldsymbol{\alpha})$ is simpler than

$$\mathscr{L}(\boldsymbol{X}^n;\boldsymbol{\alpha})=\prod_{i=1}^{n}f(\boldsymbol{x}_i;\boldsymbol{\alpha}).$$

This result also applies to (1.12) above since sufficient statistics of fixed dimension can usually be computed from sums of terms obtained from individual $\boldsymbol{x}_i$ in addition to the maximum or minimum of terms from $\boldsymbol{X}^n$ – hence these statistics can easily be couched in a form suitable for iterative estimation.

All this is very well in general but unfortunately mixture densities

$$f(\boldsymbol{x};\boldsymbol{\alpha})=\sum_{i=1}^{c}p_i g_i(\boldsymbol{x};\boldsymbol{\alpha}_i)$$

do not usually admit sufficient statistics and so the likelihood functions will not factorize in the form (1.13) (except in special cases – for example, when only one component at a time is non-zero as $\boldsymbol{\theta}$ varies). The implication is that (1.11) and (1.12) become more and more complicated as n increases.

Since general formal Bayes's procedures are computationally infeasible for mixture distributions, attempts have been made to replace them by simplifying approximations. Titterington (1976) assumes that a prior p.d.f. has been obtained from observations from known components so that initial estimates of the components' parameters can be obtained. These are then updated using observations drawn from the entire distribution. He suggests two simplifying assumptions which make such updating feasible. The first is to treat a new observation as belonging to one of the components if its probability of belonging to that component, $P(j|\boldsymbol{x})$, is 'definite' enough [e.g. if $P(j|\boldsymbol{x})$ exceeds some threshold]. The parameters of this component are then updated while those of the other components remain unchanged. His second suggestion, and the one which perhaps uses more of the available information, is to allocate fractions of each new observation to each component according to the relative sizes of $P(j|\boldsymbol{x})$ (cf. the EM algorithm). He gives an example of applying the method to a multivariate normal mixture. Of course, one drawback of all these approximations is that the results depend on the order of presentation of the observations.

Smith and Makov (1978) apply similar approximations to the special case when only the mixing proportions p_i are to be estimated. They begin with a Dirichlet prior

$$f(\boldsymbol{p}) = \frac{\Gamma[\alpha_1^{(0)} + \ldots + \alpha_c^{(0)}]}{\Gamma[\alpha_1^{(0)}] \ldots \Gamma[\alpha_c^{(0)}]} \prod_{i=1}^{c} p_i^{\alpha_i^{(0)} - 1}$$
$$\triangleq D\{\boldsymbol{p}\,; \alpha_1^{(0)}, \ldots, \alpha_c^{(0)}\}$$

and update it as

$$f(\boldsymbol{p}\,; \boldsymbol{x}_n) = D\{\boldsymbol{p}\,; \alpha_1^{(n)}, \ldots, \alpha_c^{(n)}\}$$
$$= D\{\boldsymbol{p}\,; \alpha_1^{(n-1)} + p(1\,; \boldsymbol{x}_n), \ldots, \alpha_c^{(n-1)}$$
$$+ p(c\,; \boldsymbol{x}_n)\}$$

where

$$p(i\,; \boldsymbol{x}_n) \propto g_i(\boldsymbol{x}_n)\hat{p}_i^{n-1}(\boldsymbol{x}_{n-1})$$

and the $\hat{p}_i^{(n-1)}$ is computed from the recurrence relation

$$\hat{p}_i^{(n)}(\boldsymbol{x}_n) = \hat{p}_i^{(n-1)}(\boldsymbol{x}_{n-1}) - a_{n-1}\{\hat{p}_i^{(n-1)}(\boldsymbol{x}_{n-1}) - p(i\,; \boldsymbol{x}_n)\}$$

with

$$a_{n-1} = \left[\sum_{i=1}^{c} a_i^{(0)} + n\right]^{-1}.$$

Again, a disadvantage of this 'quasi-Bayes' method is that the results depend on the order of presentation of the $\boldsymbol{x}_i$. Smith and Makov have performed a number of simulation studies and claim to have obtained a satisfactory performance.

1.4.3 *Inversion and error minimization*

In this section we outline several apparently distinct mixture parameter estimation methods which in fact have a common basis. This common basis may be expressed as follows. We begin by setting up a system of equations

$$T_j(\boldsymbol{\alpha}) = \phi_j(\boldsymbol{X}^n) \qquad j = 1, \ldots, m \tag{1.14}$$

where $\boldsymbol{\alpha}$ is the parameter vector of the mixture, $T_j(\boldsymbol{\alpha})$ is the theoretical value of a population statistic evaluated at $\boldsymbol{\alpha}$, and $\phi_j(\boldsymbol{X}^n)$ is the empirical value of this statistic, computed from the observed data $\boldsymbol{X}^n = \{\boldsymbol{x}_1, \dots, \boldsymbol{x}_n\}$. To illustrate, suppose that we have a two-component normal mixture. Then $\boldsymbol{\alpha} = (p, \mu_1, \mu_2, \sigma_1, \sigma_2)$. If we take the population mean as T_1, we have

$$T_1(\boldsymbol{\alpha}) = E(x) = p\mu_1 + (1-p)\mu_2$$

and

$$\phi_1(\boldsymbol{X}^n) = \sum_{i=1}^{n} x_i/n.$$

More conveniently, we can express the m Equations (1.14) as

$$\boldsymbol{T}(\boldsymbol{\alpha}) = \boldsymbol{\phi}(\boldsymbol{X}^n). \tag{1.15}$$

Now, in some cases it might be possible to find an m and a T_j set such that (1.15) can be inverted to give

$$\boldsymbol{\alpha} = \boldsymbol{T}^{-1}[\boldsymbol{\phi}(\boldsymbol{X}^n)], \tag{1.16}$$

an estimate of $\boldsymbol{\alpha}$. Kabir's (1968) intervals method and the various methods of moments outlined below are examples of this approach.

More generally, it will not be possible to invert (1.15), but we can define some error function

$$e(\boldsymbol{\alpha}) = e[\boldsymbol{T}(\boldsymbol{\alpha}) - \boldsymbol{\phi}(\boldsymbol{X}^n)], \tag{1.17}$$

a measure of the difference between the observed $\boldsymbol{\phi}$ and the $\boldsymbol{T}$ predicted for a particular value of $\boldsymbol{\alpha}$. We can then estimate $\boldsymbol{\alpha}$ by the value which minimizes $e(\boldsymbol{\alpha})$. Various sums of squares functions have been used for e, and these are discussed below.

We begin with the T_j set and inversion method suggested by Kabir (1968).

(a) *Kabir's intervals*

In some cases, if there are only a few parameters, (1.15) can be inverted by *ad hoc* methods, but often more structured approaches will be required. Kabir (1968) considers a fairly general method, assuming a mixture of the form

$$f(x) = \sum_{i=1}^{c} p_i g_i(x; \mu_i) \tag{1.18}$$

where $p_i \geqslant 0, \sum p_i = 1, \mu_1 < \mu_2 < \ldots < \mu_c$, and establishes a system of equations

$$\sum_{i=1}^{c} A_i \lambda_i (\mu_i)^j = \phi_j \qquad j = 0, \ldots, 2c-1. \tag{1.19}$$

The A_i here are non-zero constants, which may be functions of the μ_i, and λ_i is a strictly monotonic function of $\mu_i (i = 1, \ldots, c)$. This is a system which is linear in powers of λ_i. An example of how to derive a system such as (1.19) from (1.18) is given below.

The inversion process involves two stages: first, inverting (1.19) to express the λ_i in terms of the ϕ_j, and second, inverting $\lambda_i(\mu_i)$ to yield the μ_i. For the first stage, Kabir adopts Prony's method as follows. Suppose that we have found constants $\beta_1, \ldots, \beta_c$ such that $\lambda_1, \ldots \lambda_c$ are the roots of

$$\lambda^c - \beta_1 \lambda^{c-1} - \beta_2 \lambda^{c-2} - \ldots - \beta_{c-1}\lambda - \beta_c = 0. \tag{1.20}$$

Then, by multiplying the jth equation in (1.19) by β_{c-j} for $j = 0, \ldots, c-1$ and the cth by -1 and adding we get

$$\sum_{j=0}^{c-1} \beta_{c-j} \left(\sum_{i=1}^{c} A_i \lambda_i^j \right) - \sum_{i=1}^{c} A_i \lambda_i^c = \sum_{j=0}^{c-1} \beta_{c-j} \phi_j - \phi_c$$

which, by virtue of (1.20), simplifies to

$$\sum_{j=0}^{c-1} \phi_j \beta_{c-j} < \phi_c.$$

Similarly, if we multiply the $(j+1)$st equation in (1.19) by β_{c-j} for $j = 0, \ldots, c-1$, and the $(c+1)$st by -1 and add we get

$$\sum_{j=0}^{c-1} \phi_{j+1} \beta_{c-j} = \phi_{c+1}.$$

Continuing in this way we can set up a system of c linear equations

$$\sum_{j=0}^{c-1} \phi_{j+k} \beta_{c-j} < \phi_{c+k} \qquad k = 0, \ldots, c-1$$

or, in matrix notation

$$\boldsymbol{Z\beta} = \boldsymbol{z}$$

where

$$\boldsymbol{\beta}' = (\beta_c, \ldots, \beta_1)$$
$$\boldsymbol{z}' = (\phi_c, \ldots, \phi_{2c-1})$$

and

$$\boldsymbol{Z} = \begin{bmatrix} \phi_0 & \phi_1 \dots \phi_{c-1} \\ \phi_1 & \phi_2 \dots \phi_c \\ \cdot & \\ \cdot & \\ \cdot & \\ \phi_{c-1} & \phi_c \dots \phi_{2c-2} \end{bmatrix}$$

When the A_i are non-zero and the μ_i are distinct, $\boldsymbol{Z}$ is non-singular and may be inverted to give

$$\boldsymbol{\beta} = \boldsymbol{Z}^{-1} \boldsymbol{z}.$$

The first stage of the inversion is then completed by substituting these estimated β_i values in (1.20) and solving for the λ_i. The second stage is trivial since λ_i is a monotonic function of $\mu_i (i = 1, \dots, c)$.

To obtain estimates of the mixing proportions $p_1, \dots, p_{c-1}$, Kabir considers the first $(c - 1)$ moments of (1.18). If we substitute the $\hat{\mu}_i$, the estimates of μ_i, into the right-hand side of (1.19), we obtain another system of equations analogous to (1.15). Corresponding to the left-hand side of (1.15) we have

$$\sum_{i=1}^{c} p_i m_{ij}(\hat{\mu}_i)$$

where $m_{ij}(\hat{\mu}_i)$ is the jth moment of the ith component, computed using $\hat{\mu}_i$ in place of μ_i in the density function of this component. Corresponding to the right-hand side of (1.15) we have

$$\frac{1}{n} \sum_{i=1}^{n} x_i^j,$$

a sample estimate of the jth moment of $f(x)$. Fortunately this new system is linear in the unknown p_i, and so is more readily inverted.

Kabir considers the special case of (1.18), where g_i may be written in the form

$$g_i(x; \mu_i) = \exp[B(\mu_i)x + C(x) + D(\mu_i)]. \qquad (1.21)$$

Details of the analysis are given in Section 3.3.

To obtain Equations (1.14), Kabir assumes that all observations fall in a finite interval (a, b) which is divided into $2c$ equal sub-intervals $I_j (j = 0, \dots, 2c - 1)$. The integral of the function $f(x)/\exp C(x)$ over the jth interval then provides T_j, and ϕ_j is obtained by observing that this integral is merely the expectation of $\exp[-C(x)]$

over the jth interval. This may be estimated by

$$\phi_j = \frac{1}{n_j} \sum_{x_i \in I_j} \exp[-C(x_i)]$$

where the summation is over the n_j observations which fall in the jth interval. Kabir demonstrates that the resulting estimators of μ_i and p_i are both consistent and asymptotically normal.

(b) *Bartholomew's intervals*
Bartholomew (1959) uses a system of Equations (1.15) which is very similar to that of Kabir, except that his intervals are semi-infinite, (a_j, ∞).

(c) *Method of moments*
One of the earliest studies in which an analysis of a mixture distribution was attempted was that of Pearson (1894) who used the method of moments to estimate the five parameters of a two-component univariate normal mixture (see Section 2.3.1). Subsequently the method of moments became one of the most popular ways of estimating the parameters of a mixture distribution. One reason for this popularity is that the maximum likelihood method requires extensive and tedious calculations so that, before computers became readily available, simpler methods had to be used. This history has resulted in several applications of the method of moments to mixtures with relatively few components and such that (1.15) can be inverted by *ad hoc* methods.

The left-hand side of (1.14) takes the form

$$\int (x-\mu)^j f(x)\,\mathrm{d}x$$

where $\mu = E(x)$, the integral being analytically evaluated to give a function of the parameters of $f(x)$. The right-hand side of (1.14) is of the form

$$\frac{1}{n} \sum_{i=1}^{n} (x_i - \bar{x})^j$$

that is, a sample estimate of the left-hand side.

Examples of this procedure are given in Sections 2.3.1 and 3.3. Section 2.3.1 illustrates that even for fairly simple cases the algebra can be quite heavy. It also seems that the method will not be practicable for a large number of parameters since, as Kendall and Stuart (1963, Vol. 1, p. 234) says, 'The sampling variance of a moment depends on the moment of twice the order, i.e. becomes very large for higher moments even when n is large.' One implication of this is that the method will be of limited value for multivariate problems.

The method does have the advantage that singularities do not occur (as they do in the maximum likelihood approach where the likelihood function can be unbounded). A disadvantage of the method, and one that is common to maximum likelihood methods when attempting to find stationary values of the likelihood, is that multiple solutions can occur. Pearson (1894) suggested comparing higher-order moments for all obtained solutions and selecting the best fitting solution on this basis. Hawkins (1972) has suggested comparing each solution with the grouped raw data by χ^2 tests. We discuss the use of χ^2 as a criterion below. Note that these methods for choosing between potential solutions have moved the approach from an inversion method, as expressed by (1.16), to an error minimization method as expressed by (1.17).

Several authors have compared the efficiency of the moments method relative to the maximum likelihood method under different circumstances, and these will be considered in subsequent chapters. It seems fair to say, however, that usually the moment estimators are inefficient. Tallis and Light (1968) attempt to improve the efficiency of moment estimators (for an exponential mixture, though it seems likely that the approach will have wider applicability) by using fractional moments. The left-hand side of (1.14) takes the form

$$\int_{-\infty}^{\infty} |x|^{r_j} \, \mathrm{d}F(x, \boldsymbol{\alpha})$$

and the right-hand side

$$\frac{1}{n} \sum_{i=1}^{n} |x_i|^{r_j}$$

where the r_j are not constrained to be integers. For the case of exponential mixtures, Tallis and Light recommend r_j values which sometimes give very great improvements in efficiency. This is discussed in detail in Section 3.3.

In other cases it may be convenient to use other forms of moment. Thus in Section 4.2.1 we outline the use of factorial moments in estimating the parameters of a binomial mixture.

(d) *Moment generating function*

The preceding methods use a system of equations which can in principle be inverted to yield estimates of the parameters. Other methods set up a large number of Equations (1.14) which, if they were consistent, could be reduced to remove dependencies and, in principle, inverted. Unfortunately sampling fluctuations mean that they will not be consistent so that rather than trying to invert we minimize an error criterion as in (1.17). There is considerable freedom of choice in both $\boldsymbol{T}$ and e. One such approach is Quandt and Ramsey's (1978) generalization of the moments method (Section 2.3.6). Rather than considering a handful of low-order moments they implicitly use all the moments by means of moment generating function

$$M(\beta) = E(\exp \beta x)$$

which can be estimated by

$$\frac{1}{n} \sum_{i=1}^{n} \exp(\beta x_i).$$

For e they approximate the integrated square error so that (1.17) becomes

$$e(\boldsymbol{\alpha}) = \sum_{j=1}^{m} \left\{ E[\exp(\beta_j x)] - \sum_{i=1}^{n} \exp(\beta_j x_i)/n \right\}^2.$$

Clearly the choice of $\beta_1, \ldots, \beta_m$ is crucial.

(e) *Other difference methods*

The same sum of squares error function has been adopted by Choi and Bulgren (1968) (see also MacDonald, 1971) who use it to measure the difference between the observed and theoretical cumulative distribution function.

We have

$$e(\boldsymbol{\alpha}) = \sum_{i=1}^{n} \left\{ \sum_{j=1}^{c} p_j F[x_{(i)}, \boldsymbol{\theta}_j] - \frac{i}{n} \right\}^2$$

where $x_{(i)}$ is the ith-order statistic of the sample.

Bartlett and MacDonald (1968) suggest a similar sum of squares criterion which they express in the general form

$$\int [\mathrm{d}F(x) - \mathrm{d}F_n(x)]^2 \Big/ \mathrm{d}H(x)$$

where F is the theoretical and F_n the observed cumulative distribution function and H is some distribution function.

Binder (1978*a*) suggests using a sum of squares of differences between the theoretical and empirical characteristic functions.

These approximations to integrated squared error functions would cease to be approximate if the data were grouped and we compared the frequency distribution with areas under the theoretical curve in each group. Such a suggestion leads naturally to the use of the sum

$$e(\boldsymbol{\alpha}) = \sum_{j=1}^{m} (T_j - \phi_j)^2 / T_j$$

where T_j is the area under the theoretical curve for group j and ϕ_j is the observed proportion for that group. This is, of course, no more than the minimum χ^2 criterion. Fryer and Robertson (1972) compare this method with grouped maximum likelihood (see Section 2.3.6 – and note that grouping data prior to maximizing the likelihood function avoids the problem of singularities).

A major limitation of the methods devised for grouped data is the restriction to low dimensionality. Not only is there the problem that the number of cells increases exponentially with increasing dimensionality, but a d-dimensional multiple integral must be evaluated for each cell. There is also the problem of choosing a suitable cell size if this is not determined naturally by the data. These disadvantages to one side, however, it seems that grouped methods such as minimum χ^2 have significant advantages for $d = 1$ with censored or truncated data if there is already a natural grouping.

1.4.4 *Other methods*

Several other estimation methods have been applied to mixture distributions but usually they have been specially developed for particular component forms or for situations where additional assumptions can be made about the parameters. Where the former is the case they will be discussed in more detail in later sections. An example of the latter is Boes (1966) who assumes that only the mixing proportions are unknown. For the two-component case (Boes also considers more than two components)

$$f(x) = pg_1(x) + (1 - p)g_2(x)$$

we have, by integrating, a family of equations of the form

$$F(x) = pG_1(x) + (1 - p)G_2(x)$$

which leads to the family of estimators

$$\hat{p}_x = \frac{F(x) - G_2(x)}{G_1(x) - G_2(x)}.$$

Boes then gives necessary and sufficient conditions on G_1 and G_2 for the uniform attainment of the Cramer–Rao bound on the variance of p_x. Unfortunately these conditions impose strict limits on the permissible types of G_i, so Boes goes on to consider the less restrictive property of θ-efficiency. Apolloni and Murino (1979) also consider this problem.

A class of methods which were popular before computers became widespread is that of graphical methods (see, for example, Section 2.3.5). For mixtures of two components of known form, Cox (1966) suggests the following approach. Often the component distributions and their separation are such that by far the greater number of points in the lower tail of the observed distribution will be from just one of the components, say g_1, and the majority of points in the upper tail will be from g_2. In such a case plots on probability paper of (i) the number of observations less than x divided by np, and (ii) the number of observations greater than x divided by $n(1 - p)$, should give straight lines for x small enough and large enough. This approach can be readily applied to distributions for which simple methods of probability plotting are available, such as normal, gamma, and Weibull mixtures. Clearly to be effective the method requires well separated components or large sample sizes.

One author who has investigated one-dimensional mixture decomposition problems extensively is Medgyessi (1961). His orientation is largely towards problems of spectral decomposition though, of course, his methods can be applied in other areas. This different orientation has its consequence in different emphases. Thus, for example, for the problems he is concerned with the mixture function $f(x)$ is obtained directly (apart from measurement error) and does not need to be estimated from an observed distribution of observations. Because of these differences we will not consider his methods in detail. Having said that, it should be pointed out that his methods possess a certain mathematical elegance and so are outlined in Section 2.3.6 for the special case of normal mixtures.

Bryant and Williamson (1978) consider a different form of maximum likelihood approach, which they term classification maximum likelihood (cf. Titterington, 1976). Instead of maximizing

$$\prod_{i=1}^{n} f(\boldsymbol{x}_i; \boldsymbol{p}, \boldsymbol{\theta}) = \prod_{i=1}^{n} \sum_{j=1}^{c} p_j g_j(\boldsymbol{x}_i; \boldsymbol{\theta}_i)$$

they maximize

$$\prod_{i=1}^{n} g_{j(xi)}(\boldsymbol{x}_i; \boldsymbol{\theta})$$

where $j(\boldsymbol{x}_i) = j$ whenever observation $\boldsymbol{x}_i$ is assigned to class j. They also show that this approach is asymptotically biased.

1.4.5 *Estimating the number of components*

Up to this point the discussion of methods for estimating the mixing proportions, p_j, and the component parameters, θ_j, have been based on the assumption that c, the number of components, was known. Deciding on the number of components is a difficult problem on which relatively little work has been done – and yet it is clearly important. What work there is may be conveniently grouped into two classes: informal graphical techniques and formal hypothesis testing.

In the first class the most obvious approach is a study of the sample histogram. Unfortunately, however, unimodality of a distribution does not imply it is not a mixture so that the histogram may be deceptive. Moreover, sample distributions from single components can easily be multimodal and thus seem to imply a mixture. Other informal graphical techniques have been suggested based on probability plotting methods, and these are outlined in Chapter 5.

Turning to the class of formal hypothesis tests, we can try various c values and compare the fits of the resulting estimates using likelihood ratios or χ^2 values, for example. Several authors, as discussed in Chapter 5, have worked along these lines. Other formal tests have been suggested for particular forms of component and these also are outlined in Chapter 5. One general conclusion which can be drawn from these tests is that a theoretical reason is provided before an attempt is made to fit a mixture distribution – otherwise, by using enough components, an adequate fit can always be found, but its adequacy does not imply its meaningfulness.

1.5 Summary

This chapter has demonstrated the wide range of areas in which mixture distributions have been applied, a range which, as the recent publication date of much of the work indicates, is continuing to grow.

The definition of a mixture distribution is followed by a discussion of the important concept of identifiability, giving examples of non-identifiable distributions and necessary and sufficient conditions for identifiability to hold.

Many estimation methods have been applied to mixtures and outlines of the methods are given, beginning with the most important maximum likelihood method. Moment methods are discussed in the context of a more general approach.

CHAPTER 2

Mixtures of normal distributions

2.1 Introduction

In this chapter we consider probability density functions for a vector random variable $\boldsymbol{x}$, of dimension d, which have the following form:

$$f(\boldsymbol{x};\boldsymbol{p},\boldsymbol{\Sigma},\boldsymbol{\mu}) = \sum_{i=1}^{c} p_i g_i(\boldsymbol{x};\boldsymbol{\Sigma}_i,\boldsymbol{\mu}_i) \qquad (2.1)$$

where $\boldsymbol{p} = (p_1, p_2, \ldots, p_{c-1})$ are the $c-1$ independent mixing proportions of the mixture and are such that

$$0 < p_i < 1 \qquad p_c = 1 - \sum_{i=1}^{c-1} p_i.$$

$\boldsymbol{\Sigma} = \{\boldsymbol{\Sigma}_1, \boldsymbol{\Sigma}_2, \ldots, \boldsymbol{\Sigma}_c\}$ is a $(d \times cd)$ matrix, and $\boldsymbol{\mu} = \{\boldsymbol{\mu}_1, \boldsymbol{\mu}_2, \ldots, \boldsymbol{\mu}_c\}$ is a $(d \times c)$ matrix, in which $\boldsymbol{\mu}_i$ and $\boldsymbol{\Sigma}_i$ are the mean vector and variance–covariance matrix, respectively, of the ith component multivariate normal density, $g_i(\boldsymbol{x};\boldsymbol{\Sigma}_i,\boldsymbol{\mu}_i)$, given of course by

$$g_i(\boldsymbol{x};\boldsymbol{\Sigma}_i,\boldsymbol{\mu}_i) = (2\pi)^{-d/2} |\boldsymbol{\Sigma}_i|^{-1/2} \exp - \tfrac{1}{2}(\boldsymbol{x} - \boldsymbol{\mu}_i)' \boldsymbol{\Sigma}_i^{-1} (\boldsymbol{x} - \boldsymbol{\mu}_i). \qquad (2.2)$$

(When $d = 1$ we shall use $\sigma_1^2, \sigma_2^2, \ldots, \sigma_c^2$ for the variances of the component densities, and when $c = 2$ we shall use p and q for the mixing proportions in place of p_1 and p_2.) In the univariate case such finite normal mixture distributions have a long history, and the problem of estimating their parameters is one of the oldest estimation problems in the statistical literature. For example, Pearson (1894) attempted the estimation of the five parameters in a mixture of two normal distributions using the method of moments. The numerical solution of the resulting system of equations was, however, formidable, as it involved finding the real roots of a ninth-degree polynomial; a simpler solution of the system was given by Charlier (1906), although great computational difficulties still remained. Such difficulties led to consideration of special cases where simpler methods could be developed. For example, Charlier and Wicksell (1924) treated the two-component mixture where the two means are known, and also considered the case where the standard deviations of the components are equal. Doetsch (1928) dealt with mixtures of more than two components, again using the method of moments, and further developments were the use of cumulants instead of moments by Burrau (1934) and the use of k-statistics by Rao (1948). All such contributions offered comparatively small improvements on the basic work by Pearson, Charlier, and Wicksell.

Other early work involved the use of graphical techniques. For example, Hald (1948, 1952), Harding (1949), and Cassie (1954) used variants of normal probability plots for the estimation of the parameters in a mixture of two normal distributions. Again Oka (1954) and Bhattacharya (1967) describe methods involving plots of the logarithms of the ratio of the observed frequencies in adjacent classes.

Maximum likelihood estimation in the context of normal mixtures appears to have been first suggested by Rao (1948), who developed an iterative solution for the case of two components with equal standard deviations, based on Fisher's method of scoring. Hasselblad (1966) treated the more general case with unequal variances and more than two components, and various other workers, for example Day (1969), Wolfe (1970), Duda and Hart (1973), and Hosmer (1973) have also considered this type of estimation procedure for the parameters of normal mixture distributions.

Not surprisingly in view of the many difficulties encountered in the univariate situation, the general multivariate normal mixture density has only been considered seriously in the last two decades or so. For example, Cooper (1964) considers moment estimators for various special cases, and Wolfe (1965, 1967, 1969, 1970) in a series of papers considers maximum likelihood estimators. Day also considers the latter in the special situation where $c = 2$ and the assumption is made that both component densities have the same variance–covariance matrix.

Many of the methods mentioned above and a number of other estimation techniques which have been proposed will be discussed in detail later in this chapter, but preceding such discussion it may be of interest to consider briefly some of the descriptive properties of normal mixtures and this we shall do in the following section.

2.2 Some descriptive properties of mixtures of normal distributions

Fig. 2.1 to 2.5 show a number of examples of normal mixture distributions and indicate clearly the variety of shapes possible, and a number of authors including Harris and Smith (1949), Eisenberger (1964), Robertson and Fryer (1969), and Behboodian (1970) have considered how the parameters of a mixture determine its shape. For example, the two-component univariate normal mixture is symmetrical if

(a) $p = q$ and $\sigma_1 = \sigma_2$, or
(b) $\mu_1 = \mu_2$.

Particular interest has centred on the conditions under which such distributions are uni- rather than multimodal, and a number of results for the special case of a two-component mixture have been derived. For example, Eisenberger (1964) shows the following:

(a) If $\mu_1 = \mu_2$ then the mixture is unimodal for all $p, 0 < p < 1$.

(b) A *sufficient* condition that the mixture is unimodal for all p is that

$$(\mu_2 - \mu_1)^2 < \frac{27\sigma_1^2\sigma_2^2}{4(\sigma_1^2 + \sigma_2^2)}. \tag{2.3}$$

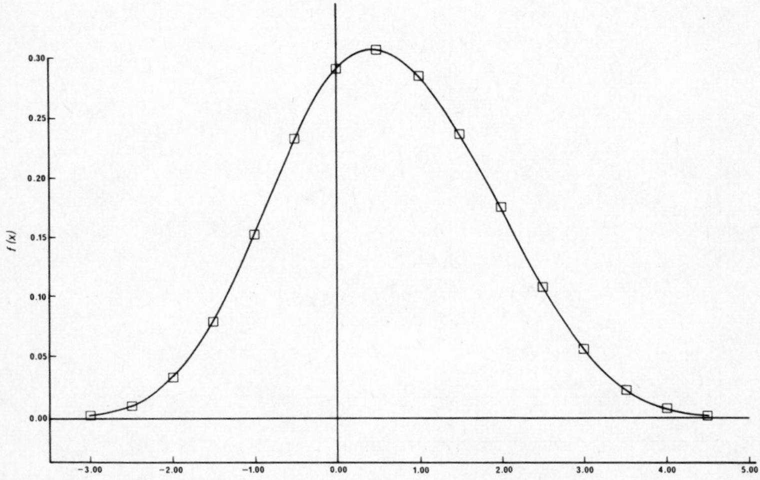

Fig. 2.1. *Mixture of two univariate normal densities with* $p = 0.3$, $\mu_1 = 0.0$, $\sigma_1 = 1.0$, $\mu_2 = 1.5$ *and* $\sigma_2 = 1.0$.

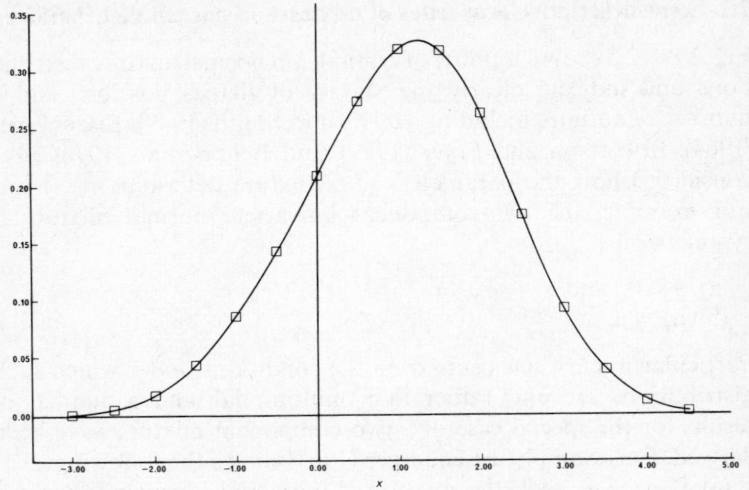

Fig. 2.2. *Mixture of two univariate normal densities with* $p = 0.6$, $\mu_1 = 0.0$, $\sigma_1 = 1.0$, $\mu_2 = 1.5$ *and* $\sigma_2 = 1.0$.

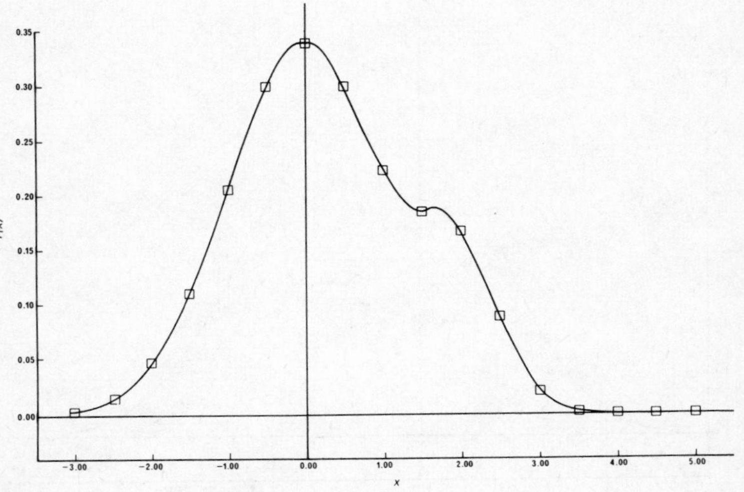

Fig. 2.3. *Mixture of two univariate normal densities with* $p = 0.85$, $\mu_1 = 0.0$, $\sigma_1 = 1.0$, $\mu_2 = 2.0$, *and* $\sigma_2 = 0.5$.

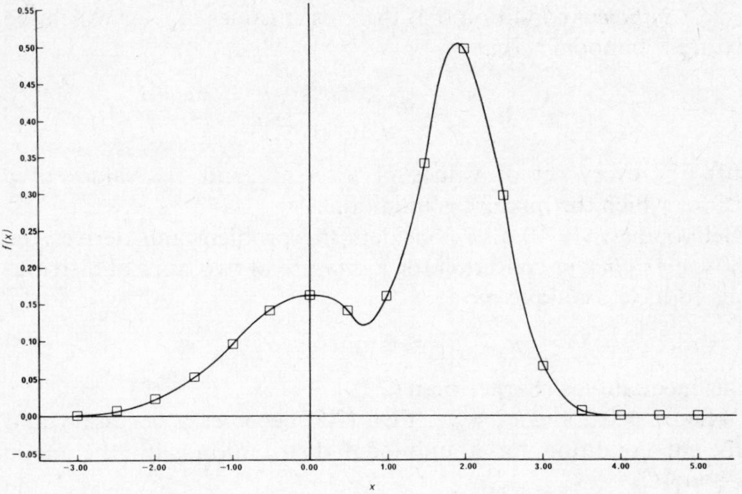

Fig. 2.4. *Mixture of two univariate normal densities with* $p = 0.4$, $\mu_1 = 0.0$, $\sigma_1 = 1.0$, $\mu_2 = 2.0$ *and* $\sigma_2 = 0.5$.

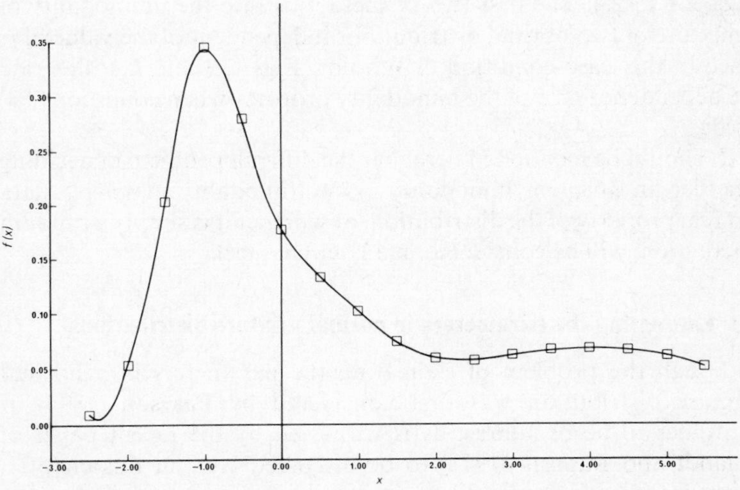

Fig. 2.5. *Mixture of three univariate normal densities with* $p_1 = 0.33$, $\mu_1 = 0.0$, $\sigma_1 = 1.0$, $p_2 = 0.33$, $\mu_2 = -1.0$, $\sigma_2 = 0.5$, $\mu_3 = 4.0$, *and* $\sigma_3 = 2.0$.

(c) A *sufficient* condition that there exist values of p for which the mixture is bimodal is that

$$(\mu_2 - \mu_1)^2 > \frac{8\sigma_1^2\sigma_2^2}{(\sigma_1^2 + \sigma_2^2)}. \tag{2.4}$$

(d) For every set of values of μ_1, μ_2, σ_1, and σ_2, values of p exist for which the mixture is unimodal.

Behboodian (1970) also considers the problem and derives the following *sufficient* condition for a mixture of two normal distributions to have a unique mode:

$$|\mu_2 - \mu_1| \leqslant 2\min(\sigma_1, \sigma_2). \tag{2.5}$$

[This inequality is sharper than (2.3).]

When $p = 0.5$ and $\sigma_1 = \sigma_2$, then (2.5) becomes a necessary and sufficient condition for a unimodal distribution with the mode $(\mu_1 + \mu_2)/2$.

Behboodian also shows that a *sufficient* condition for unimodality when $\sigma_1 = \sigma_2 = \sigma$, is given by

$$|\mu_1 - \mu_2| \leqslant 2\sigma\sqrt{(1 + |\log p - \log q|/2)} \tag{2.6}$$

(This is, of course, also a necessary condition when $p = \frac{1}{2}$).

Examples of unimodal and bimodal distributions are shown in Figs. 2.1 to 2.4. The first two of these illustrate the unimodality of a mixture of two normal distributions independent of the value of p, since in this case condition (2.3) holds. Figs. 2.3 and 2.4 illustrate the dependence on p of the bimodality property when condition (2.4) holds.

(It should be mentioned here that the difficult problem of deciding whether an *apparent* bimodality or multimodality in *sample* data is a real property of the distribution, or whether it is simply a random fluctuation, will be considered in a later chapter).

2.3 Estimating the parameters in normal mixture distributions

Although the problem of estimating the parameters in a normal mixture distribution was first considered by Pearson (1894), it continues to be of interest as is witnessed by the recent paper of Quandt and Ramsay (1978) to be discussed later in this chapter. Early attacks on the problem, including Pearson's original work, used primarily the method of moments. As a general method for estimation the method of moments lacks the optimal properties of, for example, maximum likelihood estimation, but for the separation of mixtures of normals it remains a well tried approach, and, as we shall see later, it may be useful in providing initial estimates for the iterative solution of the maximum likelihood equations. Consequently we begin our discussion of the estimation problem with a description of this particular approach.

2.3.1 *Method of moments estimation*

Consider first a two-component univariate normal mixture

$$f(x;\mu_1,\mu_2,\sigma_1,\sigma_2,p) = pg_1(x;\mu_1,\sigma_1) + (1-p)g_2(x;\mu_2,\sigma_2) \quad (2.7)$$

with g_1 and g_2 given by the univariate form of Equation (2.2), and let $x_1, x_2, \ldots, x_n$ represent our sample values. By equating the observed moments given by

$$V_r = \frac{1}{n}\sum_{i=1}^{n}(x_i - \bar{x})^r \qquad r = 0,1,\ldots,5 \quad (2.8)$$

to the theoretical moments given by

$$v_r = \int (x-\mu)^r f(x)\,\mathrm{d}x \qquad r = 0,1,\ldots,5 \quad (2.9)$$

[where $\mu = E(x)$] we can obtain the following system of five non-linear simultaneous equations, which must be solved to give estimates of the five parameters in the mixture distribution:

$$\begin{gathered} p\delta_1 + (1-p)\delta_2 = 0 \\ p(\sigma_1^2 + \delta_1^2) + (1-p)(\sigma_2^2 + \delta_2^2) = V_2 \\ p(3\delta_1\sigma_1^2 + \delta_1^3) + (1-p)(3\delta_2\sigma_2^2 + \delta_2^3) = V_3 \\ p(3\sigma_1^4 + 6\sigma_1^2\delta_1^2 + \delta_1^4) + (1-p)(3\sigma_2^4 + 6\sigma_2^2\delta_2^2 + \delta_2^4) = V_4 \\ p(15\sigma_1^4\delta_1 + 10\sigma_1^2\delta_1^3 + \delta_1^5) + (1-p)(15\sigma_2^4\delta_2 + 10\sigma_2^2\delta_2^3 + \delta_2^5) = V_5 \end{gathered} \quad (2.10)$$

where

$$\delta_k = (\mu_k - \mu) \qquad k = 1,2.$$

By some fairly tedious algebra (see Cohen, 1967), these equations may be reduced to the 'fundamental nonic', originally derived by Pearson. This takes the form

$$\sum_{i=0}^{9} a_i u^i = 0 \quad (2.11)$$

where

$$\begin{aligned} a_9 &= 24 & a_4 &= 444V_3^2k_4 - 18k_5^2 \\ a_8 &= 0 & a_3 &= 288V_3^4 - 108V_3k_4k_5 + 27k_4^3 \\ a_7 &= 84k_4 & a_2 &= -(63V_3^2k_4^2 + 72V_3^3k_5) \\ a_6 &= 36V_3^2 & a_1 &= -96V_3^4k_4 \\ a_5 &= 90k_4^2 + 72V_3k_5 & a_0 &= -24V_3^6 \end{aligned}$$

and where $k_4 = V_4 - 3V_2^2$ and $k_5 = V_5 - 10V_3V_2$ are the fourth and fifth sample cumulants. If a solution to the system of Equations (2.10) exists it may be derived from the real negative root (or roots; see later) of (2.11), as follows.

Let $\hat{u}$ be a real negative root of (2.11); then estimates of δ_1 and δ_2 are obtained as roots of the equation

$$\delta^2 - \frac{\hat{w}}{\hat{u}}\delta + \hat{u} = 0 \tag{2.12}$$

where

$$\hat{w} = \frac{-8V_3\hat{u}^3 + 3k_5\hat{u}^2 + 6V_3k_4\hat{u} + 2V_3^3}{2\hat{u}^3 + 3k_4\hat{u} + 4V_3^2}.$$

The estimates of the five parameters in the mixture may now be derived as follows:

$$\begin{aligned} \hat{\mu}_1 &= \bar{x} + \hat{\delta}_1 \\ \hat{\mu}_2 &= \bar{x} + \hat{\delta}_2 \\ \hat{\sigma}_1^2 &= \tfrac{1}{3}\hat{\delta}_1(2\hat{w}/\hat{u} - V_3/\hat{u}) + V_2 - \hat{\delta}_1^2 \\ \hat{\sigma}_2^2 &= \tfrac{1}{3}\hat{\delta}_2(2\hat{w}/\hat{u} - V_3/\hat{u}) + V_2 - \hat{\delta}_2^2 \\ \hat{p} &= \frac{\hat{\delta}_2}{\hat{\delta}_2 - \hat{\delta}_1}. \end{aligned}$$

(It should be noted that this solution is only valid for the case $\mu_1 \neq \mu_2$; the appropriate solution for the case when $\mu_1 = \mu_2$ and other special cases will be mentioned later.)

Many methods are available for finding the required roots of the ninth-degree polynomial (2.11); for example, the inclusion algorithm of Gargantini and Henrici (1967), or the technique described by Muller (1956). A difficulty which may arise is the presence of more than a single negative real root of (2.11), leading to multiple solutions of (2.10). Pearson was aware of this problem and recommended choosing the set of estimates which resulted in closest agreement between the *sixth* central moment of the sample and the corresponding moment of the fitted mixture distribution. An alternative criterion is to choose the solution with the lowest chi-squared value obtained from comparing observed and expected frequencies.

Various special cases have been considered where constraints are put on some of the parameters. Perhaps the most practically relevant is when it is assumed that the two components have equal variances. In this case only sample moments up to order four are needed and the computations simplify greatly, now involving finding the negative real roots of a cubic rather than a nonic (see Cohen, 1967, for details). Estimation by moments has also been considered in the case where the mixture is symmetric, that is when $\mu_1 = \mu_2$ or $p = \frac{1}{2}$ and $\sigma_1 = \sigma_2$, and again the calculations are greatly eased. Day (1969) and Tan and Chang (1972) have investigated the efficiency of the moment estimators and a summary of their results is given at the end of Section 2.3.2.

Example

Let us now examine how the method of moments estimation procedure performs in practice by applying it the data shown in Table 2.1. These data are the ash content of 430 peat samples given originally by Hald (1952). The sample moments for these data are also shown in Table 2.1, and these give rise to the parameter estimates shown in Table 2.2.

(It is of interest to note that Equation (2.11) for these data has two negative real roots; however, in this case the second of these leads to negative variance estimates and may therefore be disregarded. Hawkins (1972), gives an example where both negative roots lead to acceptable parameter estimates, so necessitating the use of one of the criteria mentioned above to choose between solutions.) □

A number of authors, for example Cohen (1967) and Holgersson and Jorner (1976), have considered whether Sheppard's correction should be applied in the computation of sample moments from grouped data. The later authors make the interesting point that as the correction is based on the assumption of normality it may not be applicable for normal mixtures; consequently they prefer the use of small class intervals and calculation of the moments without the correction. In many cases the parameter estimates resulting from corrected and uncorrected moments will be very similar; certainly this is the case with Hald's ash content data as shown by Table 2.2.

Table 2.1. *Ash content of* 430 *peat samples.*

Ash (%)	*Frequency*	*Ash* (%)	*Frequency*
0.25	1	6.25	35
0.75	1	6.75	43
1.25	2	7.25	48
1.75	5	7.75	45
2.25	12	8.25	35
2.75	18	8.75	26
3.25	20	9.25	17
3.75	19	9.75	13
4.25	16	10.25	9
4.75	14	10.75	4
5.25	20	11.25	2
5.75	25		

Sample moments
$V_2 = 4.84569$
$V_3 = -4.17737$
$V_4 = 58.83684$
$V_5 = -101.97396$

Table 2.2. *Estimates for the parameters of a two-component normal mixture fitted to the data of Table* 2.1.

	Component	*Proportion*	*Mean*	*Standard deviation*
(a) Moment estimates (Uncorrected sample statistics)	1	0.24	3.42	1.14
	2	0.76	7.41	1.46
(b) Moment estimates (Sheppard correction applied to sample statistics)	1	0.24	3.41	1.13
	2	0.76	7.40	1.45
(c) Maximum likelihood estimates (Taken from Hasselblad, 1966)	1	0.22	3.21	1.00
	2	0.78	7.34	1.49

Moment estimators for the two-component normal mixture distribution involve the solution of the ninth-degree polynomial (2.11). The computation involved will obviously be even more formidable if a mixture of more than two normals is considered. Even for $c = 3$, moments up to order eight are needed and the resulting system of equations is likely to be very difficult to solve with any accuracy, quite apart from the poor sampling properties of such high-order sample moments. Consequently it is unlikely that the method of moments will be of any general use in estimating the parameters in a mixture of normals with more than two-component distributions, and so we are led to consider other possible estimation procedures such as maximum likelihood.

In the multivariate situation, moment estimators have been considered by a number of authors. Perhaps the earliest were Charlier and Wicksell (1924), who consider such estimators for mixtures of two bivariate normal distributions. In the case where both variables are assumed to be uncorrelated within each distribution and also to have the same variance, Charlier and Wicksell show that by equating sample and population mixed moments of the form

$$v_{rs} = E(x_1 - \mu_1)^r (x_2 - \mu_2)^s; V_{rs} = \frac{1}{n}\Sigma(x_1 - \bar{x}_1)^r (x_2 - \bar{x}_2)^s \quad (2.13)$$

estimates of the seven parameters in the mixture may be obtained from two quadratic equations. They also show that the solution may be simplified somewhat by first rotating the system of coordinates so that the x-axis passes through the centroids of the two components.

Day (1969) also considers moment estimators for the multivariate normal mixture with two components, under the assumption that the component densities have the same variance–covariance matrix. He shows that in the multivariate situation a choice has to be made as to which third and fourth moments, or functions of these moments, to use to obtain the moment estimates. Day chooses a function of the third and fourth moments which is invariant under rotation of the sample space. By means of a simulation study he shows that moment estimators behave, in the univariate case, almost as well as maximum likelihood estimators. In two and higher dimensions, however, their sampling behaviour is very poor and consequently it would appear that they should not be used in practice in the multivariate case.

2.3.2 *Maximum likelihood estimation*

Maximum likelihood estimators are well known to have desirable asymptotic properties (see Chapter 1) and it is natural to consider the method for estimating the parameters in a mixture of normal distributions. The log of the likelihood function is given by

$$L = \sum_{i=1}^{n} \log_e f(\boldsymbol{x}_i; \boldsymbol{p}, \boldsymbol{\Sigma}, \boldsymbol{\mu}) \tag{2.14}$$

$$= \sum_{i=1}^{n} \log_e \left\{ \sum_{k=1}^{c} p_k g_k(\boldsymbol{x}_i; \boldsymbol{\Sigma}_k, \boldsymbol{\mu}_k) \right\}. \tag{2.15}$$

The maximum likelihood equations are obtained by equating the first partial derivatives of (2.15) with respect to the p_k, the elements of each matrix $\boldsymbol{\Sigma}_k$, and those of each vector $\boldsymbol{\mu}_k$, to zero. This differentiation becomes somewhat simpler if we let the independent elements of $\boldsymbol{\Sigma}_k^{-1}$ rather than $\boldsymbol{\Sigma}_k$ be the unknown parameters. Remembering that only half the off-diagonal cells of $\boldsymbol{\Sigma}_k^{-1}$ are independent and that $\sum_{k=1}^{c} p_k = 1$, we are led to the following series of equations:

$$\frac{\partial L}{\partial p_k} = \sum_{i=1}^{n} \frac{1}{f(\boldsymbol{x}_i; \boldsymbol{p}, \boldsymbol{\Sigma}, \boldsymbol{\mu})} [g_k(\boldsymbol{x}_i; \boldsymbol{\Sigma}_k, \boldsymbol{\mu}_k) - g_c(\boldsymbol{x}_i; \boldsymbol{\Sigma}_c, \boldsymbol{\mu}_c)] = 0$$

$$k = 1, 2, \ldots, c-1 \tag{2.16}$$

$$\frac{\partial L}{\partial \mu_{kl}} = \sum_{i=1}^{n} \frac{p_k g_k(\boldsymbol{x}_i, \boldsymbol{\Sigma}_k, \boldsymbol{\mu}_k)}{f(\boldsymbol{x}_i; \boldsymbol{p}, \boldsymbol{\Sigma}, \boldsymbol{\mu})} \sum_{j=1}^{d} \sigma_k^{lj}(x_{ij} - \mu_{kj}) = 0$$

$$k = 1, 2, \ldots, c; \quad l = 1, 2, \ldots, d \tag{2.17}$$

$$\frac{\partial L}{\partial \sigma_k^{lj}} = \sum_{i=1}^{n} \frac{p_k g_k(\boldsymbol{x}_i; \boldsymbol{\Sigma}_\mu, \boldsymbol{\mu}_k)}{f(\boldsymbol{x}_i; \boldsymbol{p}, \boldsymbol{\Sigma}, \boldsymbol{\mu})} (1 - \delta_{lj}/2)[\sigma_{lj}^k - (x_{il} - \mu_{kl})(x_{ij} - \mu_{kj})] = 0$$

$$k = 1, 2, \ldots, c; \quad l, j = 1, 2, \ldots, d \tag{2.18}$$

where x_{ij} $(j = 1, \ldots, d)$ are the elements of the vector $\boldsymbol{x}_i$, μ_{kj} $(j = 1, \ldots, d)$ are the elements of the vector $\boldsymbol{\mu}_k$, and σ_{ij}^k and σ_k^{ij} $(i, j = 1, \ldots, d)$ are the elements of $\boldsymbol{\Sigma}_k$ and $\boldsymbol{\Sigma}_k^{-1}$, respectively; δ_{ij} is the Kronecker delta.

As the components are assumed to exist in a fixed proportion in the mixture, we may talk about the probability that a particular sample point belongs to one of the components. If $P(s|\boldsymbol{x}_i)$ is the posterior probability that observation $\boldsymbol{x}_i$ belongs to component s, then we have

$$P(s|\boldsymbol{x}_i) = \frac{p_s g_s(\boldsymbol{x}_i; \boldsymbol{\Sigma}_s, \boldsymbol{\mu}_s)}{f(\boldsymbol{x}_i; \boldsymbol{p}, \boldsymbol{\Sigma}, \boldsymbol{\mu})}. \tag{2.19}$$

Using (2.19) the solution to Equations (2.16), (2.17), and (2.18) in terms of p_k, $\boldsymbol{\mu}_k$ and $\boldsymbol{\Sigma}_k$ may be written in the intuitively attractive form

$$\hat{p}_k = \frac{1}{n} \sum_{i=1}^{n} \hat{P}(k|\boldsymbol{x}_i) \qquad k = 1, \ldots, c-1 \tag{2.20}$$

$$\hat{\boldsymbol{\mu}}_k = \frac{1}{n\hat{p}_k} \sum_{i=1}^{n} \hat{P}(k|\boldsymbol{x}_i)\boldsymbol{x}_i \qquad k = 1, \ldots, c \tag{2.21}$$

$$\hat{\boldsymbol{\Sigma}}_k = \frac{1}{n\hat{p}_k} \sum_{i=1}^{n} \hat{P}(k|\boldsymbol{x}_i)(\boldsymbol{x}_i - \hat{\boldsymbol{\mu}}_k)(\boldsymbol{x}_i - \hat{\boldsymbol{\mu}}_k)' \qquad k = 1, \ldots, c. \tag{2.22}$$

Written in this form we can see that the maximum likelihood estimators for the parameters of a mixture of normals are closely analogous to those for estimating the parameters of a single normal distribution except that each sample point is weighted by the posterior probability (2.19). In the extreme case where $P(k|\boldsymbol{x}_i)$ is unity when $\boldsymbol{x}_i$ is from component k and zero otherwise, then $\hat{p}_k$ is simply the fraction of samples from component k, $\hat{\boldsymbol{\mu}}_k$ is the mean vector of those samples, and $\hat{\boldsymbol{\Sigma}}_k$ their variance–covariance matrix. More generally, $P(k|\boldsymbol{x}_i)$ is between zero and one, and all the samples play some role in the estimates.

Equations (2.20), (2.21), and (2.22) do not, of course, give the estimators explicitly; instead they must be solved using some type of iterative procedure. Perhaps the simplest is that suggested by Hasselblad (1966) and Wolfe (1969) which is essentially an application of the EM algorithm (see Dempster *et al.*, 1977). Initial estimates of the p_k, $\boldsymbol{\mu}_k$ and $\boldsymbol{\Sigma}_k$ are obtained by one of a variety of methods (see later), and these are then used to obtain first estimates of the $P(k|\boldsymbol{x}_i)$; these are then inserted into Equations (2.20) to (2.22) to give revised parameter estimates and the process is continued until some convergence criterion is satisfied. Such a procedure is extremely easy to implement on a computer and in the case of well separated components and good initial values appears to converge quite rapidly in both the univariate and multivariate situations. However, the results do depend upon the starting point and the possibility of multiple solutions is always present; this and other problems with the algorithm will be discussed in more detail later after some practical examples of its use have been given.

An alternative method of solving Equations (2.20) to (2.22) is to use a Newton–Raphson iterative scheme. This type of procedure has been considered for univariate normal mixtures by Hasselblad (1966) who gives the necessary second derivatives of (2.16) to (2.18). Arranging these to form an $(m \times m)$ Hessian matrix $\boldsymbol{H}$, where $m = 3c - 1$ is the number of mixture parameters to be estimated, the iteration scheme is as follows:

$$\boldsymbol{\Theta}^{(r+1)} = \boldsymbol{\Theta}^{(r)} - \boldsymbol{H}(\boldsymbol{\Theta}^{(r)})^{-1}\boldsymbol{G}(\boldsymbol{\Theta}^{(r)}) \tag{2.23}$$

where $\boldsymbol{\Theta}^{(r)}$ is the m-dimensional vector of estimates at the rth stage of the iterative procedure, $\boldsymbol{H}(\boldsymbol{\Theta}^{(r)})$ is the matrix of second derivatives evaluated at $\boldsymbol{\Theta}^{(r)}$, and $\boldsymbol{G}(\boldsymbol{\Theta}^{(r)})$ is the m-dimensional vector of first-order derivatives [given by Equations (2.16) to (2.18)] evaluated at $\boldsymbol{\Theta}^{(r)}$. Hasselblad (1966) found that the first iterative scheme always increased the likelihood, whereas Newton's method did not. On the other hand if Newton's procedure did converge, it always improved the results in fewer iterations. In general the EM procedure appeared better on smaller amounts of data with fewer grouping intervals, whilst Newton's method worked better on larger problems. This procedure appears to be impractical for the multivariate normal mixture problem, since m in this case would be $[c(d+1)(d+2)/2 - 1]$ and with, for example, $c = 3$ and $d = 5$, the Newton–Raphson algorithm would involve the repeated inversion of a (62×62) matrix.

Some further possible algorithms which may be useful in particular instances are described by Day (1969) and Ross (1970).

Before proceeding to consider some numerical examples, and to discuss the properties of maximum likelihood estimators in this context, a fundamental problem with this approach, which has been briefly mentioned in Chapter 1, needs to be examined. This problem may be illustrated by the following simple example.

Denote the likelihood function generated by a sample $x_1, x_2, \ldots, x_n$ from a two-component univariate normal mixture by $L(p, \mu_1, \mu_2, \sigma_1, \sigma_2)$; then clearly

$$L(p, x_i, \mu_2, 0, \sigma_2) = L(p, \mu_1, x_i, \sigma_1, 0) = \infty \qquad i = 1, \ldots, n$$

and so each sample point generates a singularity in the likelihood function. Similarly, any pair of sample points which are sufficiently close together will generate a local maximum, as will triplets, quadruplets, and so on which are sufficiently close. Maximum likelihood clearly breaks down and we appear to be forced to conclude that the maximum likelihood principle fails for this class of normal mixtures; consequently it seems that we must make some assumption about the parameters $\boldsymbol{\Sigma}_1, \boldsymbol{\Sigma}_2, \ldots, \boldsymbol{\Sigma}_c$ in the mixture before attempting to apply maximum likelihood. Perhaps the most natural assumption is that $\boldsymbol{\Sigma}_1 = \boldsymbol{\Sigma}_2, \ldots = \boldsymbol{\Sigma}_c = \boldsymbol{\Sigma}_A$, that is that all component distributions have the same variance–covariance matrix. If this assumption is made, Equations (2.16), (2.17), (2.20), and (2.21) remain the same but Equations (2.18) and (2.22) now become

$$\frac{\partial L}{\partial \sigma^{lj}} = \sum_{i=1}^{n} \frac{1}{f(\boldsymbol{x}_i; \boldsymbol{p}, \boldsymbol{\Sigma}, \boldsymbol{\mu})} \sum_{k=1}^{c} p_k g_k(\boldsymbol{x}_i; \boldsymbol{\Sigma}_A, \boldsymbol{\mu}_k)(1 - \delta_{lj}/2)[\sigma_{lj} - (x_{il} - \mu_{kl})(x_{ij} - \mu_{kj})] = 0 \tag{2.24}$$

$$\hat{\boldsymbol{\Sigma}}_A = \frac{1}{n} \sum_{i=1}^{n} \boldsymbol{x}_i \boldsymbol{x}_i' - \sum_{k=1}^{c} \hat{p}_k \hat{\boldsymbol{\mu}}_k \hat{\boldsymbol{\mu}}_k' \tag{2.25}$$

where σ_{ij} and σ^{ij} are the elements of the assumed common variance–covariance matrix, $\boldsymbol{\Sigma}_A$, and its inverse, respectively. The iterative procedures previously described could again be used to solve the maximum likelihood equations.

The assumption that all components have the same variance, while realistic in some situations, is obviously very restrictive, and we should perhaps consider in more detail whether a maximum likelihood approach can only be usefully employed on such a subset of normal mixture distributions. Certainly Day (1969) appears to be suggesting that this is the case. However, other authors indicate the contrary. For example, Duda and Hart (1973) find empirically that meaningful maximum likelihood solutions *can* be obtained in the unequal variances situation if attention is restricted to the largest of the finite local maxima of the likelihood function. Again simulation work by Hosmer (1973, 1974) shows that for reasonable sample sizes and initial values, the iterative maximum likelihood estimators will not converge to parameter values associated with the singularities, although these may present a problem for maximum likelihood estimation when the sample size is small and the components are not well separated. (For an example of an actual set of univariate data troubled by the singularity problem, see Murphy and Bolling, 1967.) Such findings seem to indicate that maximum likelihood will, in many cases, lead to useful estimates of the parameters in a normal mixture even when the assumption of equal variance is not made. (See Cox and Hinkley, 1974, Chapter 9, for more comments about this problem.)

The asymptotic variance–covariance matrix of the maximum likelihood estimators may be obtained from the inverse of Fisher's information matrix, that is the inverse of a matrix whose *ij*th element is given by

$$E\left[\frac{\partial \log_e f(\boldsymbol{x};\boldsymbol{\alpha})}{\partial \alpha_i}\frac{\partial \log_e f(\boldsymbol{x};\boldsymbol{\alpha})}{\partial \alpha_j}\right] \tag{2.26}$$

where the vector $\boldsymbol{\alpha} = (\alpha_1, \alpha_2, \ldots, \alpha_m)$ contains *all* the parameters of the distribution given by (2.1), so that $m = c - 1 + cd + cd(d+1)/2$. In the case of mixtures of normal distributions the elements of this matrix involve multidimensional integrals which have to be evaluated numerically. Nevertheless the asymptotic properties of the estimators have been studied by a number of authors. For example, Hill (1963) considers a univariate two-component mixture with $\sigma_1 = \sigma_2$ and gives a general power series for computing the information of the mixing proportion p. He shows that if the absolute value of $(\mu_1 - \mu_2)/\sigma$, where σ^2 is the assumed common variance, lies in the interval 0.125 to 0.250, then *at least* 1600 observations are required to ensure that the standard deviation of p is no larger than 0.1.

Behboodian (1972) also considers a two-component univariate mixture and shows that the elements of information matrix for the five parameters may be obtained by the numerical evaluation of an integral of the form

$$M_{mn}(g_i, g_j) = \int_{-\infty}^{\infty} \left(\frac{x-\mu_i}{\sigma_i}\right)^m \left(\frac{x-\mu_j}{\sigma_j}\right)^n \frac{g_i(x) g_j(x)}{f(x)} \mathrm{d}x \tag{2.27}$$

where the terms are as in Equation 2.7. For example, the first element of the information matrix, i.e. that relating to the mixing proportion p, is given by

$$I(p,p) = \frac{1}{pq}[1 - M_{00}(g_1, g_2)] \tag{2.28}$$

(where $q = 1 - p$).

Behboodian discusses various ways to evaluate the integral (2.27) and supplies a number of tables which allow the elements of the information matrix to be read directly for particular two-component mixtures. As the component densities become closer and closer to each other, Behboodian shows that the information matrix approaches a singular matrix with some diagonal elements equal to zero. The same thing happens when the parameter p tends to one or zero. Consequently he concludes that for estimating the parameters in a mixture where the two components are not well separated, or which has a mixing proportion close to zero, very large samples may be needed. (The evaluation of the information matrix for a univariate normal mixture with $c > 2$ is also fairly straightforward using one of the available numerical integration procedures, and an example is presented later in this section.)

In a recent paper, Chang (1979) considers the information matrix in the case of a two-component multivariate normal mixture with assumed common variance–covariance matrix. He shows that by suitable reparameterization, the standard errors of the maximum likelihood estimates in *any* such mixture (i.e. one with any given value of d) may always be obtained from the information matrix for the three-dimensional case. Chang further shows that the three-dimensional integrals then involved can be simplified to functions of 14 one-dimensional integrals, which may be evaluated numerically fairly simply.

Example

The univariate data shown in Table 2.3 are taken from Duda and Hart (1973). These data were generated from a two-component normal mixture with $\mu_1 = -2.0$, $\mu_2 = 2.0$, $\sigma_1 = \sigma_2 = 1.0$, and $p = 0.33$.

The results of fitting a two-component normal mixture by maximum likelihood *without* the equal variance constraint are shown in Table 2.4. (The EM algorithm was used to obtain the estimates.) In this case we see that a wide variety of initial values for the five parameters lead to the same final estimates. This will not always be the case since the maximum likelihood equations do not in general have a unique solution, and as with any iterative process there is no guarantee that the estimates associated with the *global* maximum of the likelihood will be found; instead estimates corresponding to one of perhaps many *local* maxima might result. In such cases the

Table 2.3. *Twenty-five observations from a two-component normal mixture with* $\mu_1 = -2.0$, $\mu_2 = 2.0$, $\sigma_1 = \sigma_2 = 1.0$, $p = 0.33$.

Observation	x_i	*Observation*	x_i
1	0.608	14	2.400
2	− 1.590	15	− 2.499
3	0.235	16	2.608
4	3.949	17	− 3.458
5	− 2.249	18	0.257
6	2.704	19	2.569
7	− 2.473	20	1.415
8	0.672	21	1.410
9	0.262	22	− 2.653
10	1.072	23	1.396
11	− 1.773	24	3.286
12	0.537	25	− 0.712
13	3.240		

Table 2.4. *Maximum likelihood estimates of the parameters in a two-component normal mixture fitted to the data of Table* 2.3.

	p	μ_1	μ_2	σ_1	σ_2
(i) Initial values	0.50	− 1.00	1.00	0.50	0.50
Final values	0.27	− 2.40	1.49	0.58	1.34
(ii) Initial values	0.30	− 2.00	2.00	0.50	0.50
Final values	0.27	− 2.40	1.49	0.58	1.34
(iii) Initial values	0.20	0.00	1.00	2.00	1.00
Final values	0.27	− 2.40	1.49	0.58	1.34
(iv) Moment estimates	0.37	− 1.89	1.84	0.83	1.12

solution associated with the largest of these would be accepted. For interest the moment estimates for these data are also shown in Table 2.4.

An adaptive quadrature method was used to evaluate the integrals involved in estimating the information matrix $\boldsymbol{I}$, and the following estimated covariance matrix of the maximum likelihood estimators was obtained as $(n\boldsymbol{I})^{-1}$:

Covariance matrix

1	0.0099	0.0066	0.0084	0.0049	– 0.0079
2	0.0066	0.0746	0.0258	0.0196	– 0.0224
3	0.0084	0.0258	0.1338	0.0187	– 0.0341
4	0.0049	0.0196	0.0187	0.0427	– 0.0158
5	– 0.0079	– 0.0224	– 0.0341	– 0.0158	0.0833

□

Example

The second example involves a famous set of multivariate data, namely Fisher's iris data, given in Kendall and Stuart (1963, Vol. 3). These data consist of 150 four-dimensional observations, made up of 50 observations from each of three species of iris. The four measurements taken on each plant were *sepal length*, *sepal width*,

Table 2.5. *Maximum likelihood estimation for the parameters in a three-component normal mixture on Fisher's iris data.*

	Final parameter estimates				*Initial values*			
Component 1								
$\hat{p}_1 =$	0.33				0.33			
$\hat{\boldsymbol{\mu}}_1 =$	[5.01	3.43	1.46	0.25]	[4.00	4.00	2.00	1.00]
$\hat{\boldsymbol{\Sigma}}_1 =$	0.12	0.10	0.02	0.01	1.00	0.00	0.00	0.00
	0.10	0.14	0.01	0.13	0.00	1.00	0.00	0.00
	0.02	0.01	0.03	0.01	0.00	0.00	1.00	0.00
	0.01	0.13	0.01	0.01	0.00	0.00	0.00	1.00
Component 2								
$\hat{p}_2 =$	0.30				0.33			
$\hat{\boldsymbol{\mu}}_2 =$	[5.91	2.78	4.20	1.30]	[7.00	2.00	3.00	2.00]
$\hat{\boldsymbol{\Sigma}}_2 =$	0.27	0.10	0.18	0.05	1.00	0.00	0.00	0.00
	0.10	0.09	0.09	0.04	0.00	1.00	0.00	0.00
	0.18	0.09	0.20	0.06	0.00	0.00	1.00	0.00
	0.05	0.04	0.06	0.03	0.00	0.00	0.00	1.00
Component 3								
$\hat{\boldsymbol{\mu}}_3 =$	[6.54	2.95	5.48	1.98]	[8.00	4.00	5.00	3.00]
$\hat{\boldsymbol{\Sigma}}_3 =$	0.38	0.09	0.30	0.06	1.00	0.00	0.00	0.00
	0.09	0.11	0.08	0.05	0.00	1.00	0.00	0.00
	0.30	0.08	0.32	0.07	0.00	0.00	1.00	0.00
	0.06	0.05	0.07	0.08	0.00	0.00	0.00	1.00

Procedure took 21 iterations to converge

petal length, and *petal width*. Applying the maximum likelihood estimation procedure to these data to fit a three-component, four-variate mixture gives the results shown in Table 2.5. (The known *a priori* categorization of the observations was, of course, ignored for this exercise.) This table shows parameter estimates obtained when fairly good starting values are provided for the mixing proportions and for the mean vectors. The final parameter values are extremely close to the known values for the three species of iris. However, when the starting values for the mean vectors of the three-components are altered to the following

(i)	4.0	4.0	3.0	2.0
(ii)	6.0	1.0	3.0	0.5
(iii)	7.0	4.0	3.0	4.0

and those for the other parameters left as in Table 2.5, problems arise because the algorithm converges to parameter estimates associated with a singularity. This illustrates the problem that can be encountered with singularities in the likelihood surface if no restrictions are placed on the parameters, and also indicates that the choice of starting values is likely to be far more critical in the multivariate than the univariate situation, even with components which are fairly well separated (as are the groups in Fisher's data).

□

A number of authors have investigated the properties of maximum likelihood estimators for normal mixture distributions and compared these estimators with those obtained by the method of moments and described in Section 2.3.1. For example, Hosmer (1973) uses Monte Carlo methods to investigate maximum likelihood estimators for two-component normal mixtures when $|\mu_1 - \mu_2| < 3 \min\{\sigma_1, \sigma_2\}$ and $n < 300$. He finds their sampling behaviour in general unsatisfactory, suggesting that far larger samples are needed to derive stable estimates. Again Day (1969) uses simulation to compare the two types of estimator when the two components are assumed to have equal variance, and finds that for $d = 1$ moment estimators do almost as well as maximum likelihood; for $d > 1$, however, the moment estimators perform very poorly. Tan and Chang (1972) derive the asymptotic covariance matrix of the moment estimators by utilizing the results given in Kendall and Stuart (1963, Vol. 3) for the variances and covariances of Fisher's k statistics. By comparing the elements of this matrix with the variances of the maximum likelihood estimators derived from the information matrix, they find that the method of maximum likelihood is superior to the method of moments in all cases and particularly when the separation between components is small. Hosmer (1973) investigates the maximum likelihood estimators when the data consist of a moderate sized sample from the mixed distribution and smaller samples from each of the components of the mixture, and Dick and Bowden (1973) consider the estimators when independent sample information is available from one of the populations. Both situations lead to considerable gains in efficiency over estimates computed from samples where nothing is known about which component density an observation comes from.

The above discussion indicates that whilst maximum likelihood estimation is, in most cases, far more satisfactory than the method of moments, it may itself not give accurate estimates unless the component distributions are well separated or the sample sizes very large. Also of importance, particularly in the multivariate case, is to provide the iterative estimation algorithm with good starting values for the parameters, and methods for providing such initial estimates will be briefly discussed in Section 2.3.4.

2.3.3 *Maximum likelihood estimates for grouped data*

In many cases where a mixture of univariate normal distributions is to be fitted to sample data, the observations will have previously been formed into a histogram and the raw data will not be available. Consequently we need to consider the changes that need to be made to the maximum likelihood equations given above in the case of grouped data. For this purpose let us suppose that the number of groups or classes into which the sample has been divided is k, the number of sample observations in the ith class is ϕ_i, and the upper boundary of class i is x_i. The log-likelihood of the observations is given by

$$L = \sum_{i=1}^{k} \phi_i \log_e P_i \tag{2.29}$$

where P_i is the probability of an observation falling in the ith interval, and is therefore given by

$$P_i = \int_{x_{i-1}}^{x_i} f(x)\,\mathrm{d}x \tag{2.30}$$

where $f(x)$ is the univariate form of (2.1). If the component variances are relatively large compared with the length of the intervals, then the above probabilities can be approximated by the integrand evaluated at the class midpoint, and if this is done the maximum likelihood equations for grouped data are similar in form to those given in (2.20) to (2.22), the necessary changes being that the summations are now from 1 to k, the number of classes, the observations occurring in the equations are now the class midpoints, and the frequency ϕ_i is now introduced inside the summations. In most cases this approximation will be entirely satisfactory.

Example

The data shown in Table 2.6 were originally studied by Pearson (1914), and consist of the length in micrometers of a parasitic protozoon called trypanosome. The data are grouped into intervals of 1 μm extending from 14.5 μm to 33.5 μm. The total sample size

Table 2.6. *Pearson trypanosome data.*

Length (μm)	*Frequency*
14.5–15.5	10
15.5–16.5	21
16.5–17.5	56
17.5–18.5	79
18.5–19.5	114
19.5–20.5	122
20.5–21.5	110
21.5–22.5	85
22.5–23.5	85
23.5–24.5	61
24.5–25.5	47
25.5–26.5	49
26.5–27.5	47
22.5–28.5	44
28.5–29.5	31
29.5–30.5	20
30.5–31.5	11
31.5–32.5	4
32.5–33.5	4
	1000

was 1000. Two different strains of protozoon were combined in this sample, and a mixture of two normal distributions fitted to the data by the maximum likelihood method for grouped data has the following parameter estimates:

$$\hat{p}_1 = 0.65(0.50) \qquad \hat{\mu}_1 = 19.96(20.00) \qquad \hat{\sigma}_1 = 2.15(1.00)$$
$$\hat{\mu}_2 = 26.16(26.00) \qquad \hat{\sigma}_2 = 2.76(1.00).$$

(The values in brackets are the starting values used in the maximum likelihood estimation algorithm.) The estimated standard errors (s.e.) of the parameters obtained from the estimated information matrix are

$$\text{s.e.}(\hat{p}_1) = 0.08 \qquad \text{s.e.}(\hat{\mu}_1) = 0.27 \qquad \text{s.e.}(\hat{\sigma}_1) = 0.14$$
$$\text{s.e.}(\hat{\mu}_2) = 0.68 \qquad \text{s.e.}(\hat{\sigma}_2) = 0.64.$$
□

2.3.4 *Obtaining initial parameter values for the maximum likelihood estimation algorithms*

In order to obtain initial estimates in the univariate case, Hasselblad uses a technique due to Hald (1949). This assumes that a number of 'cut-off' points are given such that nearly all the sample of the $j+1$th component lies to the right of one of these points, and some of the jth component lies to the left. Since it is assumed that the smallest cut-off point has only elements of the first sub-population to its left, the mean and variance of the first component are estimated using Hald's table of the *one-sided truncated normal distribution with known point of truncation.* Knowing these values for the first component, the estimated frequencies of the sample lying *beyond* the first cut-off point are subtracted from the originally observed frequencies and the process repeated until all c components have been separated.

For a two-component univariate mixture the moment estimators could be used to provide initial values for the maximum likelihood estimation algorithms. A further possibility in the univariate case is to use initial estimates derived from the graphical techniques to be described in the next section.

In many cases involving univariate data the choice of starting values will not be critical. However, in the multivariate situation satisfactory initial estimates are almost essential if one is to avoid misleading solutions. The example given above involving Fisher's iris data is a case in point. Perhaps the most obvious way to obtain suitable initial values for the parameters when fitting mixtures of multivariate normal distributions is to apply some form of cluster analysis to the data, and take cluster means, etc., as starting values for component mean vectors, etc. in the maximum likelihood estimation algorithm. (Cluster analysis methods are described in detail in Everitt, 1980.) For example, Wolfe (1969) uses a combination of a particular hierarchical cluster method and a 'k-means' algorithm (see MacQueen, 1967) to provide initial parameter values in his NORMIX programs for fitting multivariate normal mixtures by maximum likelihood methods. Perhaps the most suitable clustering techniques in this context are those involving minimization of trace ($\boldsymbol{W}$) and $|\boldsymbol{W}|$, where $\boldsymbol{W}$ is the pooled within clusters matrix of sums of squares and cross products. (Both these methods are described in Everitt, Chapter 3.) These methods are particularly suitable since they are, in effect, approximations to the maximum likelihood solution; see Duda and Hart (1973) and Scott and Symon (1971.)

Of course, in both the univariate and multivariate cases a further possibility for obtaining initial parameter values is that they are supplied by the investigator concerned, perhaps from *a priori* knowledge or previous work in the area.

Whilst the methods of moments and maximum likelihood are the two most common estimation techniques used in the analysis of mixtures of normal distributions, a number of other methods have been proposed and will be examined in the next two sections.

2.3.5 *Graphical estimation techniques*

Without access to a computer, both the moment and maximum likelihood procedures discussed above would be of little relevance for the routine estimation of the parameters in a normal mixture since they involve very heavy amounts of arithmetic. A consequence of this, perhaps, was the development in the 1940s and early 1950s, a period before computers were widely available, of a number of techniques intended to make the analysis of such mixture distributions a less formidable computational problem. Several of these were of a graphical nature. For example, Harding (1949) describes a method based on normal probability plots. Such plots of data from a single normal distribution result in a straight line and have long been used to obtain rough estimates of mean and variance. With data from a mixture of normals such plots lead to sigmoidal type curves which can be used to obtain estimates of the means, variances, and proportions of the component distributions. The first stage in such a procedure is the detection of the points of inflexion of the curve; Harding (1948) and Cassie (1954) determine these by eye, but as Fig. 1 in the latter author's paper indicates, this can be an extremely subjective process. Fowlkes (1979) suggests a more objective technique for determining the points of inflexion based on fitting a modified logistic curve, but it appears unlikely, unless the components are very well separated, that such probability plots will lead to very accurate parameter estimates. (They may, however, be useful for other purposes – see Chapter 5.)

A more satisfactory graphical technique for grouped data is perhaps that suggested by Bhattacharya (1967). He suggests a plot $\log \phi_{i+1}/\phi_i$ against x_i, where ϕ_i and ϕ_{i+1} are the observed frequencies of adjacent classes i and $i+1$, and x_i is the mid-point of class i. Bhattacharya shows that such a plot should lead to a series of approximately straight lines with negative slopes, each line corresponding to an area where one component dominates. He further demonstrates that estimates of the parameters μ_k and σ_k^2 of the kth component normal distribution can be obtained from the plot as follows:

$$\hat{\mu}_k = \lambda_k + w/2 \tag{2.31}$$

$$\sigma_k^2 = w \cot(\alpha_k) - w^2/12 \tag{2.32}$$

where α_k is the angle between the kth straight line and the negative direction of the x-axis, λ_k is the x-intercept of the line, and w is the class width. After the means and variances have been determined in this way, Bhattacharya suggests several methods which might be used to determine the mixing proportions. For details, readers are referred to the original paper.

Example

To demonstrate how Bhattacharya's method works in practice it was applied to the data shown in Table 2.7. These data were generated from a four-component normal mixture with the following characteristics:

Component 1:	$p_1 = 0.25$	$\mu_1 = 2.0$	$\sigma_1 = 0.5$
Component 2:	$p_2 = 0.25$	$\mu_2 = 5.0$	$\sigma_2 = 0.5$
Component 3:	$p_3 = 0.25$	$\mu_3 = 9.0$	$\sigma_3 = 1.0$
Component 4:	$p_4 = 0.25$	$\mu_4 = 15.0$	$\sigma_4 = 2.0$

Table 2.7. *Data generated from a four-component normal mixture distribution.*

x_i	ϕ_i	ϕ_{i+1}/ϕ_i	$\log_e \phi_{i+1}/\phi_i$
0.5	2	11.50	2.44
1.5	23	0.96	− 0.04
2.5	22	0.05	− 3.00
3.5	1	37.00	3.61
4.5	37	0.76	− 0.27
5.5	28	0.04	− 3.22
6.5	1	6.00	1.79
7.5	6	2.83	1.04
8.5	17	0.94	− 0.06
9.5	16	0.31	− 1.17
10.5	5	0.40	− 0.92
11.5	2	1.50	0.41
12.5	3	2.00	0.69
13.5	6	1.00	0.00
14.5	6	2.50	0.92
15.5	15	0.40	− 0.92
16.5	6	0.33	− 1.11
17.5	2	1.00	0.00
18.5	2		
	200		

Applying the plotting procedure to these data leads to Fig. 2.6. The first three lines are easy to place, and lead to the following estimates of means and standard deviations:

$$1: \hat{\mu}_1 = 2.1 \qquad \hat{\sigma}_1 = 0.65$$

$$2: \hat{\mu}_2 = 4.9 \qquad \hat{\sigma}_2 = 0.45$$

$$3: \hat{\mu}_3 = 9.3 \qquad \hat{\sigma}_3 = 1.15$$

The placing of the fourth line is more problematical, although it is fairly clear that a fourth component is present. Fitting the line to the points with the largest associated frequency as shown gives the following estimates for component four:

$$4: \hat{\mu}_4 = 14.5 \qquad \hat{\sigma}_4 = 1.27$$

For comparison the maximum likelihood estimators for a four-component mixture fitted to these data are as follows:

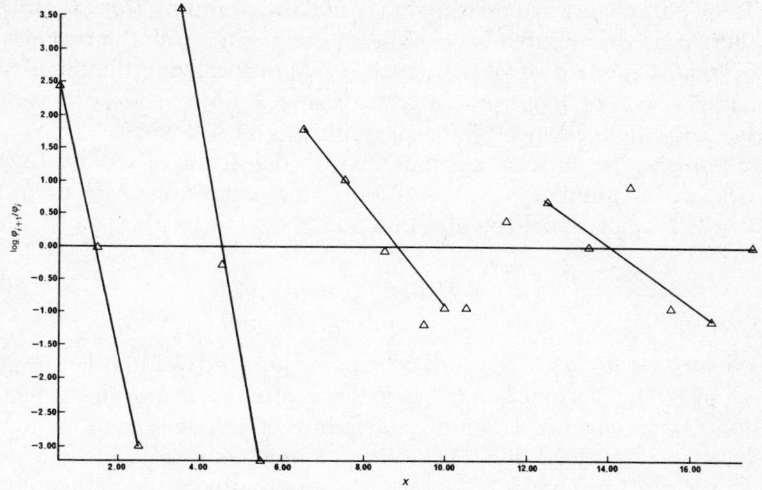

Fig. 2.6. *Bhattacharya's method on the data in Table 2.7.*

Component 1:	$\hat{\mu}_1 = 1.94$	$\hat{\sigma}_1 = 0.59$
Component 2:	$\hat{\mu}_2 = 4.92$	$\hat{\sigma}_2 = 0.52$
Component 3:	$\hat{\mu}_3 = 8.96$	$\hat{\sigma}_3 = 0.96$
Component 4:	$\hat{\mu}_4 = 15.09$	$\hat{\sigma}_4 = 1.60$

(The initial values provided were the parameter values used to generate the data.) □

Graphical techniques such as those described above are unlikely to give accurate parameter estimates unless the components in the mixture are well separated. However, they may be extremely useful in an initial examination of data since they have one great advantage over methods such as moments and maximum likelihood, namely that they function without *a priori* knowledge of the number of components in the mixture. Consequently they may prove helpful in indicating the number of components, information which is necessary before applying either of the other two methods just mentioned. This aspect of the use of graphical techniques will be taken up in more detail in Chapter 5.

2.3.6 *Other estimation methods*

A number of other methods have been proposed for finding estimates of the parameters in a mixture of normal distributions. For example, there is a considerable body of literature arising from the proposal originally made by Doetsch (1928, 1936) of localizing the different components of a normal mixture using Fourier transformation methods. The essence of the procedure is as follows.

Suppose we have a normal mixture density as given by (2.1). Choose a number $\lambda > 0$ satisfying the condition $\lambda < \min\{\sigma_i\}$ $(i = 1, 2, \ldots, c)$, and form the function

$$f^*(x; \boldsymbol{\theta}^*, \mathbf{p}) = \sum_{i=1}^{c} p_i g_i(x; \boldsymbol{\theta}_i^*) \tag{2.33}$$

where $\boldsymbol{\theta}^* = (\boldsymbol{\theta}_1^*, \boldsymbol{\theta}_2^* \ldots \boldsymbol{\theta}_c^*)$ with $\boldsymbol{\theta}_i^* = [\mu_i, \sqrt{(\sigma_i^2 - \lambda^2)}]$ and other terms as in (2.1). The function f^* is again a mixture of normal density functions, but the standard deviation of a component is now $\sqrt{(\sigma_i^2 - \lambda^2)}$ instead of σ_i. The other parameters remain unchanged. In the plots of f and f^*, the latter has more obvious local maxima, and if the value of λ is sufficiently close to the smallest deviation then the mean of the component with this standard deviation is the x value where the function has its absolute maximum.

To derive f^* from f (which we take as the observed frequency distribution), a Fourier transformation is employed along with an initial value of λ which may be changed as the procedure progresses. The abscissa of the absolute maximum of f^* is now used as an estimate of μ_1, the mean of the component with the smallest standard deviation, and as an approximate estimate of σ_1 we may take λ. Gregor (1969) then shows how to estimate p_1. Next we form the function $\bar{f}$ as

$$\bar{f}(x) = f(x) - \hat{p}_1 g_1(x; \hat{\mu}_1, \hat{\sigma}_1) \tag{2.34}$$

and repeat the process with $\bar{f}$ and a value of λ satisfying $\sigma_1 < \lambda < \sigma_2$, where σ_2 is the next smallest standard deviation. This is repeated until the residual frequencies of the form (2.34) are less than some error term.

Detailed discussion of this approach is given in Medgyessi (1961), and an algorithm is provided by Gregor (1969). There appears to have been no investigations of the sampling properties of this method or attempts to compare it with statistically more conventional techniques such as moments or maximum likelihood.

In a recent paper, Quandt and Ramsey (1978) describe a technique for estimating the parameters in a mixture of two univariate normal distributions, which uses the moment generating function of the mixture, that is

$$E\exp(\beta x) = p\exp(\mu_1\beta + \sigma_1^2\beta^2/2) + (1-p)\exp(\mu_2\beta + \sigma_2^2\beta^2/2). \tag{2.35}$$

For any given value of β, say β_j, the moment generating function may be estimated by

$$\hat{E}\exp(\beta_j x) = \frac{1}{n}\sum_{i=1}^{n}\exp(\beta_j x_i) \tag{2.36}$$

and Quandt and Ramsey propose estimating the parameters of the mixture by minimizing the expression

$$S = \sum_{j=1}^{k}[\hat{E}\exp(\beta_j x) - E\exp(\beta_j x)]^2 \tag{2.37}$$

that is, the sum of squares of differences between the sample moment generating function and the theoretical moment generating function, for values of β, $\beta_1, \beta_2, \ldots, \beta_k$. Obviously crucial in this respect is choice of the values $\beta_1, \ldots, \beta_k$. Quandt and Ramsey discuss this issue in some detail and suggest that very large or very small values of β should be avoided and that k should be not less than five. Their choice in the examples they give was $k = 5$ with β values -0.2, -0.1, 0.1, 0.2, 0.3. Minimization of (2.37) may be achieved by one of the variety of minimization algorithms now available. Quandt and Ramsey employ the method due to Davidon and Fletcher-Powell. The results of a simulation study described in the paper show that this moment generating function method compares very favourably with the method of moments, and some further simulation results given by Hosmer (1978) appear to indicate that it also outperforms maximum likelihood in many situations. However, this last observation is perhaps thrown into doubt by some further simulation work reported by Kumar, Nicklin, and Paulson (1979), who find that the method of maximum likelihood completely dominates the moment generating function procedure. They suggest that the differences in the various studies are produced because Quandt and Ramsey and Hosmer use the true parameter values as initial estimates in the algorithm used to minimize (2.37). Kumar and co-workers also indicate that the moment generator estimates are very sensitive to the location of the β_j.

Some results of the present authors appear to confirm that Kumar *et al.* are correct. Applying maximum likelihood and the moment generating procedure to mixtures of *more* than two univariate normal components (the extension of the moment generating function technique to the case $c > 2$ is relatively trivial) shows that the latter is far more sensitive to choice of starting values than is maximum likelihood. Tables 2.8 and 2.9 show a number of the simulation results.

Table 2.8. *A comparison of maximum likelihood with Quandt and Ramsey's estimation method.*

	Maximum likelihood	*Quandt and Ramsey*
p_1		
MSE	0.006 43	0.002 87
Var	0.006 40	0.002 84
Ave	0.323 84	0.335 23
p_2		
MSE	0.010 56	0.003 22
Var	0.010 55	0.002 88
Ave	0.333 44	0.348 56
μ_1		
MSE	0.125 69	0.177 58
Var	0.125 69	0.177 28
Ave	0.000 65	0.017 13
μ_2		
MSE	0.164 38	0.336 71
Var	0.164 32	0.318 26
Ave	4.007 35	4.135 85
μ_3		
MSE	0.108 97	0.056 82
Var	0.107 91	0.056 63
Ave	8.032 61	8.013 98
σ_1		
MSE	0.062 19	0.149 94
Var	0.060 37	0.136 33
Ave	0.957 37	1.116 65
σ_2		
MSE	0.142 03	0.267 65
Var	0.137 05	0.216 69
Ave	0.929 38	1.225 74

Table 2.8. (*Contd.*)

	Maximum likelihood	*Quandt and Ramsey*
σ_3		
MSE	0.074 29	0.077 30
Var	0.068 63	0.073 44
Ave	0.925 13	1.062 07

Data generated from a three-component mixture with parameter values $p_1 = p_2 = 0.33, \mu_1 = 0.0, \mu_2 = 4.0, \mu_3 = 8.0, \sigma_1 = \sigma_2 = \sigma_3 = 1.0$, and sample size 50.

Starting values used here are the true parameter values.

The results given are the mean square errors (MSE), the variance, and the average of the parameter estimates of 100 samples.

Values of β_j used in this and Table 2.9 were -0.5, -0.45, -0.4, -0.35, -0.3, -0.25, -0.15, -0.1, 0.1, 0.15, 0.20, 0.25, 0.30, 0.35, 0.40, 0.45, 0.50.

Table 2.9. *A comparison of maximum likelihood with Quandt and Ramsey's estimation method.*

	Maximum likelihood	*Quandt and Ramsey*
p_1		
MSE	0.005 86	0.017 69
Var	0.005 84	0.015 78
Ave	0.325 52	0.286 29
p_2		
MSE	0.016 88	0.050 00
Var	0.014 18	0.034 24
Ave	0.278 06	0.204 48
μ_1		
MSE	0.202 09	13.804 83
Var	0.201 92	8.497 32
Ave	-0.013 16	2.303 81
μ_2		
MSE	0.287 93	9.582 48
Var	0.247 50	9.580 63
Ave	3.798 95	3.956 96

Table 2.9. (*Contd.*)

	Maximum likelihood	*Quandt and Ramsey*
μ_3		
MSE	0.600 89	5.169 99
Var	0.490 64	1.102 50
Ave	7.667 69	5.983 20
σ_1		
MSE	0.085 49	15.190 40
Var	0.084 29	5.365 11
Ave	0.965 31	4.134 53
σ_2		
MSE	0.267 27	23.344 16
Var	0.211 20	21.738 09
Ave	0.763 20	2.267 31
σ_3		
MSE	0.269 44	13.472 44
Var	0.238 34	3.218 61
Ave	1.173 65	4.202 16

Data generated from same distribution as used in Table 2.7, with same sample size. Starting values used were as follows: $p_1 = 0.2$, $p_2 = 0.3$, $\mu_1 = 2.0$, $\mu_2 = 3.0$, $\mu_3 = 5.0$, $\sigma_1 = 2.0$, $\sigma_2 = 0.5$, $\sigma_3 = 2.0$.
The results give the mean square error (MSE), the variance, and the average of the parameter estimates for 100 samples.

Such results tend to indicate that whilst the estimation of the parameters in a mixture via the moment generating function procedure is an interesting approach, it is, in general, less satisfactory than estimation via maximum likelihood. However, it may prove useful in situations where maximum likelihood methods fail, as found by Hosmer (1978) in a slightly different context.

Of course, any measure of the difference between the empirical and theoretical distributions could be used for estimation purposes, for example, minimum χ^2, and Fryer and Robertson (1972) describe a detailed investigation of the bias and accuracy of moment estimates, maximum likelihood estimates, and minimum χ^2 estimates; they find that in terms of bias the latter seem to be slightly better than grouped maximum likelihood, but that the difference between the two procedures is seldom large. Again the estimation technique for mixture distribution developed by Choi and Bulgren (1968) which involves the minimization of the sum of squares of differences between the theoretical and empirical, sample based, distribution function of the mixture, could be used for estimating the parameters in a univariate normal mixture. (This technique is described in detail in Chapter 1.) A simulation study of the method in this particular case is reported in the paper and shows that reasonable estimates are obtained with 25 observations when $c = 2$ and the components are one standard deviation apart, and with 100 observations when $c = 5$ and the components are again each separated by one standard deviation. (The assumption is made that each component distribution has the same variance.)

A number of other estimation procedures for normal mixture densities arise from their application in pattern recognition 'unsupervised learning' research. For example, Young and Coraluppi (1976) describe a stochastic estimation algorithm, and Chien and Fu (1967) also describe some examples of stochastic approximation to estimate what is basically a two-component mixture.

2.4 Summary

The problem of estimating the parameters in a normal mixture distribution is one of the oldest estimation problems in the statistical literature. Despite this the problem still attracts a great deal of attention which probably reflects its difficulty and the lack of a completely satisfactory solution. The estimation methods discussed in this chapter should, however, provide reasonable parameter estimates in many situations provided the sample size is large enough. In all the discussions in this chapter it has been assumed that the number of components in the mixture is known, perhaps based on theoretical knowledge derived from the application at hand. In many situations, however, the investigator may wish to estimate the number of components using the sample data, and this interesting and difficult problem will be considered in Chapter 5.

CHAPTER 3

Mixtures of exponential and other continuous distributions

3.1 Exponential mixtures

Partly as a consequence of its mathematical properties and partly as a consequence of its practical importance, the single continuous distribution which has received the most attention is the normal distribution. Furthermore, as the preceding chapter illustrates, this is also true when considering particular forms for the components of mixture distributions. However, there are circumstances when the normal distribution is clearly inappropriate and other distributions have been studied. Of these the most important as a component of a mixture is the exponential distribution.

Single exponential distributions are encountered in, amongst other places, the analysis of failure data, where the probability density function of failure time can often be closely approximated by

$$f(x) = \frac{1}{\mu} \exp(-x/\mu) \qquad x \geqslant 0; \mu > 0.$$

Thus, for example, one might be concerned about the mean time between failures of a machine in a factory, or the length of time one can expect a complex system to function without failing. 'Failure' here can, of course, be interpreted in a perfectly general way as an event – the probability density function for the times of scores in a game may be exponentially distributed.

When one considers that failures may arise for a number of different reasons it seems hardly surprising that a superposition of exponential densities – a mixture – might provide a better description of the failure properties. Thus, for example, the causes of electronic valve failures have been grouped into three categories: gaseous defects, mechanical defects, and normal deterioration of the cathode. Note also that Davis (1952) says, 'Most systems fail from a combination of human and mechanical causes.' Again it is important to remember that 'failure' can be interpreted generally – Thomas (1966) fits an exponential mixture to the discharges of the brain's neurons.

The general form of a finite exponential mixture is

$$f(x) = \sum_{i=1}^{c} p_i \frac{1}{\mu_i} \exp(-x/\mu_i) \qquad x \geqslant 0 \tag{3.1}$$
$$= 0 \text{ otherwise}$$

where

$$\sum_{i=1}^{c} p_i = 1$$
$$p_i > 0 \qquad i = 1, \ldots, c$$
$$\mu_i > 0 \qquad i = 1, \ldots, c.$$

One of the complications presented by failure data is that the observed samples are often censored or truncated. Thus, for example, it may well not be feasible to test a set of items until they all fail because that might take too long. It is thus necessary for the analysis to take the non-failed items into account. It would be incorrect to exclude them since the resulting estimates, ignoring the largest observations, would be underestimating the μ_i parameters. Possible complications due to non-identifiability fortunately do not arise since, as Teicher (1961) has shown, finite mixtures of exponentials are identifiable.

The practical importance suggested by the above examples is complemented by theoretical and mathematical importance. The fact that the Laplace transform of an exponential mixture is a rational algebraic function has implications in renewal theory and the theory of stochastic processes. Stretching the definition of mixture slightly to allow negative p_i yields the useful property that any density on $(0, \infty)$ can be approximated arbitrarily closely by an expression of form (3.1). This is discussed further in Section 3.3, where conditions on the p_i and μ_i in such a 'mixture' are given for $f(x)$ to be a p.d.f. Finally, exponential mixtures truncated on the left are also exponential mixtures.

3.2 Estimating exponential mixture parameters

3.2.1 *The method of moments and generalizations*

Rider (1961) has applied the method of moments to the decomposition of a mixture of two exponential distributions:

$$f(x) = p\frac{1}{\mu_1}\exp(-x/\mu_1) + (1-p)\frac{1}{\mu_2}\exp(-x/\mu_2). \qquad (3.2)$$

The way we shall present the method will make the relation with subsequent generalizations more obvious. From (3.2) we can write the 0th, 1st, 2nd and 3rd moments about zero as

$$1 = p + (1-p) \qquad (3.3a)$$

$$\bar{x} = p\mu_1 + (1-p)\mu_2 \qquad (3.3b)$$

$$\frac{m_2}{2} = p\mu_1^2 + (1-p)\mu_2^2 \qquad (3.3c)$$

$$\frac{m_3}{6} = p\mu_1^3 + (1-p)\mu_2^3 \qquad (3.3d)$$

and for convenience we set

$$1 = T_0 \qquad \bar{x} = T_1 \qquad \frac{m_2}{2} = T_2 \qquad \frac{m_3}{6} = T_3.$$

Now, multiplying (3.3*a*) by some constant α_2, (3.3*b*) by some constant α_1, (3.3*c*) by -1, and adding, we get

$$p(-\mu_1^2 + \alpha_1\mu_1 + \alpha_2) + (1-p)(-\mu_2^2 + \alpha_1\mu_2 + \alpha_2)$$
$$= \alpha_2 T_0 + \alpha_1 T_1 - T_2.$$

Similarly, if we multiply (3.3*b*) by α_2, (3.3*c*) by α_1, and (3.3*d*) by -1, and add, we get

$$p(-\mu_1^3 + \alpha_1\mu_1^2 + \alpha_2\mu_1) + (1-p)(-\mu_2^3 + \alpha_1\mu_2^2 + \alpha_2\mu_2)$$
$$= \alpha_2 T_1 + \alpha_1 T_2 - T_3.$$

Thus, if we choose α_1 and α_2 such that μ_1 and μ_2 are the roots of $(\mu^2 - \alpha_1\mu - \alpha_2 = 0)$, we obtain

$$\alpha_2 T_0 + \alpha_1 T_1 = T_2$$

and

$$\alpha_2 T_1 + \alpha_1 T_2 = T_3$$

from which

$$\left.\begin{aligned} \alpha_1 &= (T_1T_2 - T_0T_3)/(T_1^2 - T_0T_2) \\ \text{and} \quad \alpha_2 &= (T_1T_3 - T_2^2)/(T_1^2 - T_0T_2). \end{aligned}\right\} \qquad (3.4)$$

Thus, we estimate the α_i from (3.4) using the observed T_i values, and then we estimate the μ_i from

$$\mu_1, \mu_2 = \frac{\alpha_1}{2} \mp \tfrac{1}{2}(\alpha_1^2 + 4\alpha_2)^{1/2}.$$

From $\hat{\mu}_1$ and $\hat{\mu}_2$ and Equation (3.3*b*) we can estimate $\hat{p}$.

Example

We wish to decompose the following set of observations into a mixture of two exponential components:

0.16	0.61	0.62
0.23	1.51	0.50
0.71	2.03	0.01
0.09	0.41	0.08

We have $\bar{x} = 0.58, m_2 = 0.34$, and $m_3 = 0.28$, and hence $T_0 = 1$, $T_1 = 0.58, T_2 = 0.68$, and $T_3 = 1.68$. Thus

$$\hat{\alpha}_1 = \frac{0.58 \times 0.68 - 1 \times 1.68}{0.58 \times 0.58 - 1 \times 0.68} = 3.74$$

$$\hat{\alpha}_2 = \frac{0.58 \times 1.68 - 0.68 \times 0.68}{0.58 \times 0.58 - 1 \times 0.68} = -1.49.$$

Thus

$$\hat{\mu}_1 = \frac{\hat{\alpha}_1}{2} - \tfrac{1}{2}(\hat{\alpha}_1^2 + 4\hat{\alpha}_2)^{1/2} = 0.45$$

$$\hat{\mu}_2 = \frac{\hat{\alpha}_1}{2} + \tfrac{1}{2}(\hat{\alpha}_1^2 + 4\hat{\alpha}_2)^{1/2} = 3.29$$

and

$$\hat{p} = \frac{\bar{x} - \hat{\mu}_2}{\hat{\mu}_1 - \hat{\mu}_2} = 0.95.$$

□

Since we require $\hat{\mu}_1, \hat{\mu}_2$ real and $\hat{\mu}_1, \hat{\mu}_2 > 0$ we can put certain conditions on the $\hat{\alpha}_1$. For $\hat{\mu}_1, \hat{\mu}_2$ real we need

$$\hat{\alpha}_1^2 > -4\hat{\alpha}_2.$$

For $\hat{\mu}_1, \hat{\mu}_2 > 0$ we need

$$\hat{\alpha}_1 > 0, \hat{\alpha}_2 < 0.$$

Now, for $\hat{\alpha}_1 > 0$ we need

either $(T_1 T_2 > T_0 T_3 \text{ and } T_1^2 > T_0 T_2)$

or $(T_1 T_2 < T_0 T_3 \text{ and } T_1^2 < T_0 T_2)$.

That is,

$$\text{either} \quad \left(\frac{T_0}{T_1} < \frac{T_2}{T_3} \text{ and } \frac{T_0}{T_1} < \frac{T_1}{T_2}\right) \tag{3.5}$$

$$\text{or} \quad \left(\frac{T_0}{T_1} > \frac{T_2}{T_3} \text{ and } \frac{T_0}{T_1} > \frac{T_1}{T_2}\right). \tag{3.6}$$

Similarly, from $\hat{\alpha}_2 < 0$ we need

$$\text{either} \quad \left(\frac{T_1}{T_2} > \frac{T_2}{T_3} \text{ and } \frac{T_0}{T_1} > \frac{T_1}{T_2}\right) \tag{3.7}$$

$$\text{or} \quad \left(\frac{T_1}{T_2} < \frac{T_2}{T_3} \text{ and } \frac{T_0}{T_1} < \frac{T_1}{T_2}\right). \tag{3.8}$$

We can combine these [noting that (3.5) and (3.7) cannot be simultaneously true, and (3.6) and (3.8) cannot be simultaneously true] to require that either [from (3.5) and (3.8)]

$$\left(\frac{T_0}{T_1} < \frac{T_1}{T_2} \text{ and } \frac{T_1}{T_2} < \frac{T_2}{T_3} \text{ and } \frac{T_0}{T_1} < \frac{T_2}{T_3}\right)$$

or [from (3.6) and (3.7)]

$$\left(\frac{T_0}{T_1} > \frac{T_1}{T_2} \text{ and } \frac{T_1}{T_2} > \frac{T_2}{T_3} \text{ and } \frac{T_0}{T_1} > \frac{T_2}{T_3}\right).$$

That is, we cannot have $\hat{\mu}_1, \hat{\mu}_2 > 0$ unless the sequence

$$\frac{T_0}{T_1}, \frac{T_1}{T_2}, \frac{T_2}{T_3}$$

is monotonic (increasing or decreasing). This permits a quick check on whether the method will work.

Rider (1961) states that if $\mu_1 \neq \mu_2$ then $\hat{\mu}_1, \hat{\mu}_2$, and $\hat{p}$ are consistent estimators such that

$$P(\hat{\mu}_1 > 0,\ \hat{\mu}_2 > 0,\ 0 \leqslant \hat{p} \leqslant 1) \to 1 \text{ and } n \to \infty.$$

For $\mu_1 = \mu_2$, however, the estimators are not consistent.

For the case of p known, Rider gives [from (3.3*b*) and (3.3*c*)]

$$\hat{\mu}_1 = \bar{x} \pm \left[\frac{1-p}{2p}(m_2 - 2\bar{x}^2)\right]^{1/2}$$

(where the + is used if $\mu_1 \geqslant \mu_2$ and the − otherwise) and

$$\hat{\mu}_2 = \bar{x} \mp \left\{\frac{p}{2-2p}(m_2 - 2\bar{x}^2)\right\}^{1/2}$$

(where the − is used if $\mu_1 \geqslant \mu_2$ and the + otherwise). The first pair of estimators ($\hat{\mu}_1$ using +, $\hat{\mu}_2$ using −) is consistent when $\mu_1 \geqslant \mu_2$, and the second pair when $\mu_1 \leqslant \mu_2$ (and note that here the case $\mu_1 = \mu_2$ is handled). However, the probability that $\hat{\mu}_1$ and $\hat{\mu}_2$ are real does not approach 1 as $n \to \infty$. Obviously we will seldom know when $\mu_1 > \mu_2$ or $\mu_1 < \mu_2$. As Rider remarks: 'This admittedly is a real shortcoming of the method'.

The moment Equations (3.3) are one way to set up a system relating sample statistics [left-hand sides of (3.3)] to analytic expressions [right-hand sides of (3.3)] involving the parameters to be estimated. Kabir (1968) (see also Chapter 1) has expressed this approach in a more general way and has developed a different special case. Again we shall consider a two-component mixture.

We shall begin by assuming that all of the observations have fallen in a finite interval $(0, b)$. (This requirement might be a disadvantage in some circumstances as the discussion of truncation above illustrates.) We can divide the interval into four equal subintervals:

$$\begin{aligned} I_0 &= (0, h) \\ I_1 &= (h, 2h) \\ I_2 &= (2h, 3h) \\ I_3 &= (3h, 4h) = (3h, b). \end{aligned}$$

Defining

$$T_j = \int_{jh}^{(j+1)h} f(x)\,\mathrm{d}x \qquad j = 0, \ldots, 3 \tag{3.9}$$

gives

$$T_j = \sum_{i=1}^{2} k_i \int_{jh}^{(j+1)h} \beta_i^x\,\mathrm{d}x = \sum_{i=1}^{2} \frac{k_i(\beta_i^h - 1)}{\log \beta_i} \beta_i^{jh}$$

where

$$k_i = p_i/\mu_i \text{ and } \beta_i = \exp(-1/\mu_i). \tag{3.10}$$

For convenience we can rewrite this as

$$T_j = \sum_{i=1}^{2} A_i \lambda_i^j \qquad j = 0, \ldots, 3 \tag{3.11}$$

with

$$A_i = k_i(\beta_i^h - 1)/\log \beta_i \text{ and } \lambda_i = \beta_i^h. \tag{3.12}$$

[System (3.11) serves in a role analogous to (3.3).]

Now, it is apparent that if we can invert the system of Equations (3.11), expressing the λ_i in terms of the T_j, then, via (3.12) and (3.10), we can express the μ_i in terms of the T_j. The T_j can be estimated [from (3.9)] as

$$\hat{T}_j = n_j/n,$$

n_j being the number of observations which falls in the interval I_j.

To invert (3.11) let us write the equations out in full:

$$T_0 = A_1 \lambda_1^0 + A_2 \lambda_2^0 \tag{3.11a}$$

$$T_1 = A_1 \lambda_1^1 + A_2 \lambda_2^1 \tag{3.11b}$$

$$T_2 = A_1 \lambda_1^2 + A_2 \lambda_2^2 \tag{3.11c}$$

$$T_3 = A_1 \lambda_1^3 + A_2 \lambda_2^3 \tag{3.11d}$$

These equations are now in exactly the same form as (3.3) and can be inverted by the same method: we write an equation ($\lambda^2 - \alpha_1 \lambda - \alpha_2 = 0$) which has λ_1 and λ_2 as its roots and [by multiplying (3.11a) by α_2, etc.] we arrive at

$$\left.\begin{aligned} \alpha_1 &= (T_1 T_2 - T_0 T_3)/(T_1^2 - T_0 T_2) \\ \text{and} \quad & \\ \alpha_2 &= (T_1 T_3 - T_2^2)/(T_1^2 - T_0 T_2) \end{aligned}\right\} \tag{3.13}$$

yielding

$$\lambda_1, \lambda_2 = \frac{\alpha_1}{2} \mp \tfrac{1}{2}(\alpha_1^2 + 4\alpha_2)^{1/2}. \tag{3.14}$$

The weights p and $(1 - p)$ are calculated from the first moment

$$\bar{x} = \hat{p}\hat{\mu}_1 + (1 - \hat{p})\hat{\mu}_2,$$

that is,

$$\hat{p} = \frac{\bar{x}_1 - \hat{\mu}_2}{\hat{\mu}_1 - \hat{\mu}_2}.$$

[Note that since in (3.13) the denominators, n, of the T_j cancel out it is sufficient to let $T_j = n_j$, the number of observations falling in the jth interval.]

Example
A set of observations have a mean value of $\bar{x} = 0.92$ and the following frequency distribution:

Interval	*Number*
0–2	200
2–4	9
4–6	2
6–8	1

From Equations (3.13) we get

$$\hat{\alpha}_1 = \frac{9 \times 2 - 200 \times 1}{9 \times 9 - 200 \times 2} = \frac{-182}{-319} = 0.5705$$

$$\hat{\alpha}_2 = \frac{9 \times 1 - 2 \times 2}{9 \times 9 - 200 \times 2} = \frac{5}{-319} = -0.0157$$

giving

$$\hat{\lambda}_1 = \frac{0.5705}{2} - \tfrac{1}{2}(0.5705^2 - 4 \times 0.0157)^{1/2}$$
$$= 0.02899$$

$$\hat{\lambda}_2 = \frac{0.5705}{2} + \tfrac{1}{2}(0.5705^2 - 4 \times 0.0157)^{1/2}$$
$$= 0.54151.$$

Hence from (3.12) and (3.10)

$$\hat{\mu}_1 = -2/\log_e \hat{\lambda}_1 = 0.56$$
$$\hat{\mu}_2 = -2/\log_e \hat{\lambda}_2 = 3.26$$

and

$$\hat{p} = \frac{0.92 - 3.26}{0.56 - 3.26} = 0.87.$$ □

The requirements that the μ_1 should exist and be greater than zero allow us to impose restrictions on the T_j just as in the moments case. We have the general relationship

$$\mu_i = -h/\log_e \lambda_i \qquad i = 1,2$$

so that

(i) for μ_i to exist $\lambda_i > 0 (i = 1, 2)$;

(ii) for $\mu_i > 0$ we need $\lambda_i < 1$ $(i = 1, 2)$.

Just as with the basic moments approach, condition (i) leads to the condition that the sequence

$$\frac{T_0}{T_1}, \frac{T_1}{T_2}, \frac{T_2}{T_3}$$

should be monotonic. Now, from Equation (3.14) we can see that condition (ii) will certainly not be satisfied if $\hat{\alpha}_1 > 2$. Thus we also require $\hat{\alpha}_1 \leqslant 2$ (necessary, but not sufficient). This condition allied to the necessity of the root in (3.14) being real, leads to

$$\hat{\alpha}_2 > \frac{-\hat{\alpha}_1^2}{4} > \frac{-4}{4} = -1.$$

The method of moments and the method presented above follow the general approach outlined in Chapter 1. For c classes we set up a system of equations

$$T_j = \sum_{i=1}^{c} A_i \lambda_i^j$$

where the T_j are sample statistics and the λ_i are known (monotonic) functions of the mixture parameters μ_i. Inversion of this system expresses the λ_i in terms of the T_j, and inversion of the function relating λ_i and μ_i yields μ_i. The weights $p_i (i = 1, \ldots, c-1)$ can be found from further expressions relating (for example) the mixture sample moments to estimates of the component population moments.

Let us consider c component mixtures of general exponential distributions of the form

$$\begin{cases} \exp[B(\mu)x + C(x) + D(\mu)] & x \in (a, b) \text{ a given interval} \\ 0 \text{ otherwise} \end{cases}$$

where $B(\mu)$ is strictly monotonic with a continuous first derivative (this distribution includes the binomial, negative binomial, and Poisson, which will be considered in the next chapter, as special cases, as well as the exponential distribution).

We can express the mixture

$$f(x) = \sum_{i=1}^{c} p_i \exp[B(\mu_i)x + C(x) + D(\mu_i)]$$

as

$$t(x) = \sum_{i=1}^{c} k_i \beta_i^x \qquad x \in (a,b)$$

where

$$t(x) = f(x) \exp[-C(x)]$$
$$k_i = p_i \exp[D(\mu_i)]$$
$$\beta_i = \exp B(\mu_i).$$

An interval (a,b) is selected which includes all the observations, and this is divided into $2c$ equal subintervals $I_0, \ldots, I_{2c-1}$ by $a = a_0, a_1, \ldots, a_{2c} = b$ such that

$$a_{j+1} - a_j = h, \qquad j = 0, \ldots, 2c - 1.$$

Now let

$$T_j = \int_{a_j}^{a_{j+1}} t(x)\,\mathrm{d}x \qquad j = 0, \ldots, 2c - 1$$

which gives

$$T_j = \sum_{i=1}^{c} \frac{k_i}{\log \beta_i} (\beta_i^{a_0 + h} - \beta_i^{a_0}) \beta_i^{jh}$$

i.e.

$$T_j = \sum_{i=1}^{c} A_i \lambda_i^j \qquad j = 0, 1, \ldots, 2c - 1$$

where

$$A_i = \frac{k_i}{\log \beta_i} (\beta_i^{a_0 + h} - \beta_i^{a_0}) \text{ and } \lambda = \beta_i^h,$$

giving a general system of equations of the type required.

Kabir (1968) then shows that

$$\hat{T}_j = \frac{1}{n} \sum_x [-C(x)]$$

[where the summation is over the $x_i (i = 1, \ldots, n)$ falling in subinterval I_j] is a consistent estimator of T_j. For the exponential case this becomes simply

$$\hat{T}_j = n_j / n$$

where n_j is the number of observations falling in I_j. He further proves that the resulting $\hat{\mu}_i$ and $\hat{p}_i$ are consistent and asymptotically normal.

Bartholomew (1959) and Ashton (1971) use a system of equations which is very similar to Kabir's system, except that instead of T_j being the proportion falling in a finite interval $[jh, (j+1)h]$ they use P_j, the proportion of observations with a value greater than jh.

The standard method of moments presented above has also been extended in an interesting way by Tallis and Light (1968). Instead of using a system of N equations of the form

$$\frac{1}{n}\sum_{j=1}^{n} x_j^a = \int_{-\infty}^{\infty} x^a f(x, \boldsymbol{\theta})\,\mathrm{d}x$$

they suggest the use of fractional a. If we define

$$S_{a_i} = \frac{1}{n}\sum_{j=1}^{n} |x_j|^{a_i} \qquad i = 1, \ldots, N$$

and

$$m_{a_i}(\boldsymbol{\theta}) = \int_{-\infty}^{\infty} |x|^{a_i} f(x, \boldsymbol{\theta})\,\mathrm{d}x \qquad i = 1, \ldots, N$$

we have the system

$$\boldsymbol{m}(\boldsymbol{\theta}) = \boldsymbol{S}. \tag{3.15}$$

An iterative correction to be applied in estimating $\boldsymbol{\theta}$ is obtained by expanding

$$\boldsymbol{S} - \boldsymbol{m}(\boldsymbol{\theta}) = H(\boldsymbol{\theta}_0)\boldsymbol{\delta}$$

where $H(\boldsymbol{\theta})$ is the matrix with ijth element $\partial m_{a_i}(\boldsymbol{\theta})/\partial\theta_j$ and $\boldsymbol{\theta}_0$ is an initial estimate for $\boldsymbol{\theta}$. If $|H(\boldsymbol{\theta}_0)| \neq 0$ we can invert H to give

$$\boldsymbol{\delta} = H(\boldsymbol{\theta}_0)^{-1}[\boldsymbol{S} - \boldsymbol{m}(\boldsymbol{\theta})].$$

The iterative solution of (3.15) then follows the steps

$$\begin{cases} \boldsymbol{\delta} = H(\boldsymbol{\theta}_k)^{-1}[\boldsymbol{S} - \boldsymbol{m}(\boldsymbol{\theta}_k)] \\ \boldsymbol{\theta}_{k+1} = \boldsymbol{\theta}_k + \boldsymbol{\delta}. \end{cases}$$

So far this is nothing but a different way of writing and solving the equations of the standard moments method. However, now let us consider the efficiency of the resulting estimates.

From Tallis and Light (1968) the large sample variance of the estimator $\hat{\boldsymbol{\theta}}$ of $\boldsymbol{\theta}$ is given by

$$V(\hat{\boldsymbol{\theta}}) = H(\boldsymbol{\theta})^{-1} V(\boldsymbol{S}) H(\boldsymbol{\theta})'^{-1}$$

where $V(\boldsymbol{S})$ (the covariance matrix of $\boldsymbol{S}$) has ijth element

$$[m_{a_i+a_j}(\theta) - m_{a_i}(\theta)m_{a_j}(\theta)]/n.$$

With the two-component exponential mixture

$$f(x) = p\frac{1}{\mu_1}\exp(-x/\mu_1) + (1-p)\frac{1}{\mu_2}\exp(-x/\mu_2)$$

we have

$$m_{a_i}(\theta) = E(|x|^{a_i}) = \Gamma(a_i+1)[p\mu_1^{a_i} + (1-p)\mu_2^{a_i}].$$

Table 3.1. *Suitable combinations of* a_2 *and* a_3 $(a_1 = 1)$ *for various values of b and p.*

		b					
p		1.5	2	3	4	5	10*
0.1	a_2	2.25	2.00	1.50	1.50	0.75	0.75
	a_3	2.75	2.25	1.75	1.75	1.50	1.25
0.2	a_2	2.00	1.75	1.25	0.75	0.75	0.75
	a_3	2.50	2.00	1.50	1.50	1.25	1.25
0.3	a_2	2.00	1.50	1.25	0.75	0.75	0.50
	a_3	2.25	1.75	1.50	1.25	1.25	0.75
0.4	a_2	1.75	1.25	0.75	0.75	0.50	0.50
	a_3	2.00	1.75	1.50	1.25	1.25	0.75
0.5	a_2	1.75	1.25	0.75	0.75	0.50	0.50
	a_3	2.00	1.50	1.25	1.25	0.75	0.75
0.6	a_2	1.50	1.25	0.75	0.50	0.50	0.25
	a_3	2.00	1.50	1.25	1.25	0.75	0.50
0.7	a_2	1.50	1.25	0.75	0.50	0.50	0.25
	a_3	1.75	1.50	1.25	0.75	0.75	0.50
0.8	a_2	1.50	1.25	0.50	0.50	0.50	0.25
	a_3	1.75	1.50	1.25	0.75	0.75	0.50
0.9	a_2	1.50	0.75	0.50	0.50	0.25	0.25
	a_3	1.75	1.50	1.25	0.75	0.75	0.50

*For $b > 10$ and all θ_3, $a_2 = 0.25$ and $a_3 = 0.50$ are satisfactory.

Using $|V(\hat{\boldsymbol{\theta}})|$ and the information matrix $I(\boldsymbol{\theta})$, Tallis and Light obtain $|V(\hat{\boldsymbol{\theta}})|^{-1}|I(\boldsymbol{\theta})|^{-1}$ to compare the generalized moments estimator with the maximum likelihood estimator to give the efficiency of this generalized method of moments. For $a_1 = 1, a_2 = 2, a_3 = 3$ (that is, the usual method of moments) they give a table of efficiency values for various values of $b = \mu_1/\mu_2$ and p. It is quite striking from this table that the efficiency falls off rapidly with increasing b and p (for example, with $b = 4$ and $p = 0.3$ the efficiency is 0.138 and for $b = 10$ and $p = 0.9$ the efficiency is 0.002). This discovery led Tallis and Light to investigate different a_2 and a_3 (they keep a_1 constant at 1) and they obtained a table of a_2, a_3 values for different b, p combinations which give the greatest efficiency. This is reproduced here as Table 3.1. Large improvements can result by choosing good values for a_2 and a_3 (for example, $b = 4$ and $p = 0.3$ leads to $a_2 = 0.75$ and $a_3 = 1.25$ with efficiency 0.671; $b = 10$, $p = 0.9$ leads to $a_2 = 0.25, a_3 = 0.50$ and efficiency 0.522). Only if $b < 1.5$ and $p < 0.3$ is there little difference between the efficiencies resulting from the different a_2, a_3 values.

For a sample censored at X, giving r observations less than X and $(n - r)$ of unknown value greater than X, we can use the same procedure if we replace $H(\boldsymbol{\theta})$ by

$$H(\boldsymbol{\theta}) = \begin{bmatrix} H_1(\boldsymbol{\theta})' \\ H_2(\boldsymbol{\theta})' \\ H_3(\boldsymbol{\theta})' \end{bmatrix}$$

where

$$H_i(\boldsymbol{\theta}) = \begin{bmatrix} -pa_i\mu_1^{a_i+1}\displaystyle\int_0^{X/\mu_1} t^{a_i}\exp(-t)\,dt \\ (1-p)a_i\mu_2^{a_i+1}\displaystyle\int_0^{X/\mu_2} t^{a_i}\exp(-t)\,dt \\ \mu_1^{a_i}\displaystyle\int_0^{X/\mu_1} t^{a_i}\exp(-t)\,dt - \mu_2^{a_i}\int_0^{X/\mu_2} t^{a_i}\exp(-t)\,dt \\ + X^{a_i}[\exp(-X/\mu_1) - \exp(-X/\mu_2)] \end{bmatrix}$$

and use

$$S_{a_i} = \left[\sum_{j=1}^{r} x_j^{a_i} + (n-r)X^{a_i}\right]\bigg/ n.$$

Joffe (1964) also considers fractional moments, though for a different reason. He is concerned with Sichel's (1957) model for the distribution of the diameter x of dust particles in mines:

$$f(x) = pA\exp(-\sqrt{x/\mu_1}) + (1-p)B\exp(-\sqrt{x/\mu_2})$$

where A and B are normalizing constants

$$A = \frac{\exp(\sqrt{\alpha/\mu_1})}{2\mu_1(\mu_1 + \sqrt{\alpha})}$$

$$B = \frac{\exp(\sqrt{\alpha/\mu_1})}{2\mu_2(\mu_2 + \sqrt{\alpha})}$$

and α is the known (and fixed) size of the smallest particles observed.

From this Joffe sets up the half-moment equations

$$S_{a/2} = \hat{p}m_{a/2}(\hat{\mu}_1) + (1-\hat{p})m_{a/2}(\hat{\mu}_2) \qquad a = 1, 2, 3 \tag{3.16}$$

where

$$S_{a/2} = \frac{1}{n}\sum_{j=1}^{n} x_j^{a/2}$$

$$m_{a/2}(\mu) = \frac{\exp(\sqrt{\alpha/\mu})}{2\mu(\mu + \sqrt{\alpha})}\int_a^\infty x^{\alpha/2}\exp(-\sqrt{x/\mu})\mathrm{d}x.$$

(The integral can be evaluated by substituting $x = u^2$ followed by integration by parts.)

From (3.16) Joffe derives

$$\hat{\mu}_2 = \frac{R - \hat{p}\hat{\mu}_1 Q}{P - \hat{p}Q} \tag{3.17}$$

where

$$P = 4S_1 - 3\sqrt{\alpha}\;\; S_{1/2}$$

$$Q = 4m_1(\hat{\mu}_1) - 3\sqrt{\alpha}m_{1/2}(\hat{\mu}_1)$$

$$R = S_{3/2} - \sqrt{\alpha}S_1$$

and

$$\hat{p} = \frac{PU - VR}{Q(U - V\hat{\mu}_1) - T(R - P\hat{\mu}_1)} \tag{3.18}$$

where

$$U = S_1 - \sqrt{\alpha}S_{1/2}$$

$$V = 3S_{1/2} - 2\sqrt{\alpha}$$

$$T = 3m_{1/2}(\hat{\mu}_1) - 2\sqrt{\alpha}.$$

System (3.16) can then be solved numerically by searching for a value of $\hat{\mu}_1$ [and via (3.18) $\hat{p}$, and (3.17) $\hat{\mu}_2$] which satisfies (3.16).

Joffe also suggests evaluating the asymptotic variances and covariances as follows. For any $\hat{p}, \hat{\mu}_1, \hat{\mu}_2$ values

$$n \operatorname{cov}(S_{r/2}, S_{s/2}) = m_{(r+s)/2} - m_{r/2} m_{s/2} \qquad r \leqslant s = 1, 2, 3 \quad (3.19)$$

can be evaluated.

Furthermore, combinations of linear expansions of the half-moments about (p, μ_1, μ_2) lead to six linear equations of the form

$$\operatorname{cov}(S_{r/2}, S_{s/2}) = \left(\frac{\partial}{\partial \hat{\mu}_1} S_{r/2} \frac{\partial}{\partial \hat{\mu}_1} S_{s/2}\right) \operatorname{var} \hat{\mu}_1$$

$$+ \left(\frac{\partial}{\partial \hat{\mu}_1} S_{r/2} \frac{\partial}{\partial \hat{\mu}_2} S_{s/2} + \frac{\partial}{\partial \hat{\mu}_1} S_{s/2} \frac{\partial}{\partial \hat{\mu}_2} S_{r/2}\right) \operatorname{cov}(\hat{\mu}_1, \hat{\mu}_2)$$

$$+ \ldots . \qquad r \leqslant s = 1, 2, 3. \qquad (3.20)$$

Combining (3.19) and (3.20) yields asymptotic variances and covariances for $\hat{\mu}_1, \hat{\mu}_2$, and $\hat{p}$.

3.2.2 *Maximum likelihood*

The basic estimation methods outlined in Chapter 1 can be applied directly, without attempting any analytic simplification. Thus, for example, the general log-likelihood function is

$$L = \sum_{j=1}^{n} \log \sum_{i=1}^{c} p_i \frac{1}{\mu_i} \exp(-x_j/\mu_i)$$

and we can seek the $\boldsymbol{p}$ and $\boldsymbol{\mu}$ which maximize L. As in earlier chapters, a simple iterative scheme can be devised. Deriving the normal equations from L and applying a little algebra yields the expressions

$$\hat{p}_i = \frac{1}{n} \sum_{j=1}^{n} \hat{\mathrm{P}}(i|x_j) \qquad (3.21)$$

and

$$\hat{\mu}_i = \frac{1}{n\hat{p}_i} \sum_{j=1}^{n} \hat{\mathrm{P}}(i|x_j) x_j \qquad (3.22)$$

where $\hat{\mathrm{P}}(i|x_j)$ is the estimate of the posterior probability that x_j comes from class i obtained by substituting parameter estimates in

$$\mathrm{P}(i|x_j) = \frac{p_i g_i(x_i; \mu_i)}{f(x_i; \mu)}. \qquad (3.23)$$

Equations (3.21), (3.22), and (3.23) permit a straightforward iterative solution as outlined in Chapter 1. Hasselblad (1969) outlines the details of this algebra for components following the general exponential distribution

$$g_i(x) = H(x)C(\theta_{1i}, \ldots, \theta_{ri}) \exp[\theta_{1i} T_1(x) + \ldots + \theta_{ri} T_r(x)].$$

Apart from the negative exponential distribution considered here, the Poisson and binomial distributions (Chapter 4) are also special cases of this $g_i(x)$.

This approach assumes that we take measurements on n objects and obtain a result for all n. That is, it requires that the time to 'failure' of all of our original sample of objects should be known. As mentioned in the preceding section, incorrect estimates of the parameters will result if this straightforward method is applied to a subsample selected by choosing only those objects which fail before a certain time.

We can handle such truncation in the following way. Given a sample of n objects, the probability that r will fail by time X is

$$\frac{n!}{r!(n-r)!} F(X)^r [1 - F(X)]^{n-r}$$

where F is the cumulative distribution function of f. The conditional density of obtaining the ordered observations $x_1, \ldots, x_r$ given r and that $x_r \leqslant X$ is

$$\frac{r! \prod_{j=1}^{r} f(x_j)}{[F(X)]^r}.$$

The likelihood function is thus

$$\mathscr{L} = \frac{n!}{(n-r)!} [1 - F(X)]^{n-r} \prod_{j=1}^{r} f(x_j).$$

Now

$$F(X) = \int_0^X \sum_{i=1}^{c} p_i \frac{1}{\mu_i} \exp(-x/\mu_i) \mathrm{d}x = 1 - \sum_{i=1}^{c} p_i \exp(-X/\mu_i)$$

giving

$$\mathscr{L} = \frac{n!}{(n-r)!} \left(\sum_{i=1}^{c} p_i \exp(-X/\mu_i) \right)^{n-r} \prod_{j=r}^{r} f(x_j).$$

Application of brute force function maximization by computer is in many ways unsatisfying. It gives no insight into the problem and no indication of conditions under which the estimates are unreliable. For these reasons it is valuable to consider alternative methods which do permit analytic solution, though only for special cases. Many of these methods were developed before computers became readily available.

Mendenhall and Hader (1958) use a maximum likelihood approach for the case when the cause of failure is (retrospectively) known and the distribution is truncated. This means that one can study the failed items and determine to which category they belong, but one cannot tell to which category the non-failed items belong. For a two-component mixture with n items under study, the probability that r_1 will fail due to cause 1, r_2 will fail due to cause 2, and $n - r_1 - r_2$ will survive at time X when testing is stopped, is

$$\frac{n!}{r_1!r_2!(n-r_1-r_2)!}[p_1 F_1(X)]^{r_1}[p_2 F_2(X)]^{r_2}[G(X)]^{n-r_1-r_2}$$

where $G(X) = 1 - p_1 F_1(X) - p_2 F_2(X)$. The conditional density of obtaining the ordered observations $x_{i1}, \ldots, x_{ir_i}$ given r_i and $x_{ij} \leqslant X (i = 1, 2; j = 1, \ldots, r_i)$ is

$$\frac{r! \prod_{j=1}^{r_i} f_i(x_{ij})}{[F_i(X)]^{r_i}}.$$

The likelihood is thus

$$\mathscr{L} = \frac{n!}{(n-r_1-r_2)!}[G(X)]^{n-r_1-r_2} p_1^{r_1} p_2^{r_2} \prod_{j=1}^{r_1} f_1(x_{1j}) \prod_{j=1}^{r_2} f_2(x_{2j}).$$

Equating the derivatives of $L = \log_e \mathscr{L}$ with respect to μ_1, μ_2, and p_1 to zero gives

$$\begin{aligned} \hat{p}_1 &= \frac{r_1}{n} + \frac{\hat{k}(n-r_1-r_2)}{n} \\ \hat{\mu}_1 &= \bar{x}_1 + \frac{\hat{k}(n-r_1-r_2)X}{r_1} \\ \hat{\mu}_2 &= \bar{x}_2 + \frac{(1-\hat{k})(n-r_1-r_2)X}{r_2} \end{aligned} \tag{3.24}$$

where

$$\bar{x}_i = \sum_{j=1}^{r_i} x_{ij}/r_i \qquad i = 1,2$$

and

$$\hat{k} = \left[1 + \frac{\hat{p}_2}{\hat{p}_1}\exp\left(\frac{X}{\hat{\mu}_1} - \frac{X}{\hat{\mu}_2}\right)\right]^{-1}. \tag{3.25}$$

Substituting (3.24) into (3.25) gives a straightforward univariate problem (note that $0 \leqslant \hat{k} \leqslant 1$).

Example

To illustrate this procedure consider the data in Table 3.2 (from Davis, 1952). Table 3.2(a) is the failure distribution for an indicator valve and Table 3.2(b) is the failure distribution for a transmitter valve, both used in aircraft radar sets. Until failure occurs one does not know which valve will fail – that is, one does not know the category of failure. The total number of tests carried out was 1003

Table 3.2. *Failure distributions for electronic valves* (*Davis*, 1952).

(a) *Indicator valves*

Time (*h*)	*Observed distribution*
0–100	29
100–200	22
200–300	12
300–400	10
400–600	10
600–800	9
	92

(b) *Transmitter valves*

Time (*h*)	*Observed distribution*
0–40	166
40–80	151
80–120	132
120–160	98
160–200	73
200–240	45
240–280	53
280–320	40
320–360	23
360–400	26
400–440	24
440–480	9
480–520	9
520–560	8
560–600	9
600–700	17
700–800	8
	891

(c) *Indicator valves*

Time (*h*)	*Expected distribution*
0–100	26.72
100–200	19.62
200–300	14.40
300–400	10.57
400–600	13.44
600–800	7.25
	92.00

and of these 20 had not failed by the time the testing was terminated. Thus $n = 1003$ and $r_1 + r_2 = 983$. From the tables we have $r_1 = 891$ and $r_2 = 92$. Thus

$$\hat{p}_1 = \frac{891}{1003} + \frac{\hat{k}(1003 - 983)}{1003} = 0.8883 + 0.0199\,\hat{k}$$

$$\hat{\mu}_1 = 169.6184 + \frac{\hat{k}(1003 - 983)}{891} \times 800$$

$$= 169.6184 + 17.9573\,\hat{k}$$

$$\hat{\mu}_2 = 245.1086 + \frac{(1 - \hat{k})(1003 - 983)}{92} \times 800$$

$$= 419.0216 - 173.9130\,\hat{k}$$

and

$$\hat{k} = \left[1 + \frac{0.1117 - 0.0199\,\hat{k}}{0.8883 + 0.0199\,\hat{k}} \times \exp\left(\frac{800}{169.6184 + 17.9573\,\hat{k}} - \frac{800}{419.0216 - 173.9130\,\hat{k}}\right)\right]^{-1}.$$

(Note: in calculating $\bar{x}_1$ and $\bar{x}_2$ we have taken the failure times as occurring at the mid-points of the intervals.)

This final equation is satisfied by $\hat{k} = 0.55$, giving

$$\hat{p}_1 = 0.90$$
$$\hat{\mu}_1 = 179.50$$
$$\hat{\mu}_2 = 323.38.$$

□

Table 3.2(c) gives the expected values for the indicator valve failures using this μ_2 value.

The maximum likelihood estimator for the μ parameter of a single exponential distribution is

$$\hat{\mu} = (\text{total observed life})/r$$

$$= \left[\sum_{j=1}^{r} x_j + (n - r)X\right] \Big/ r.$$

Comparing this with (3.24) we see that the mixture estimators follow the same form: the $\bar{x}_i$ are calculated from the failed objects which are known to come from class i; a proportion of the $(n - r)X$ part of the total observed life is assigned to each class. (In fact, since

$$\frac{\hat{p}_1 \exp(-X/\hat{\mu}_1)}{\hat{p}_1 \exp(-X/\hat{\mu}_1) + \hat{p}_2 \exp(-X/\hat{\mu}_2)} = \hat{k}$$

we are assigning a proportion which appears intuitively reasonable.)

In some cases it is known that $\mu_1 \leqslant \mu_2$ or $\mu_1 \geqslant \mu_2$. Without loss of generality let us assume $\mu_1 \leqslant \mu_2$. If we know this and yet the two-class maximum likelihood method above has produced $\hat{\mu}_1 > \hat{\mu}_2$, it would be reasonable to choose a common estimate $\hat{\mu} = \hat{\mu}_1 = \hat{\mu}_2$. In this case the problem reduces to the single population discussed above with

$$\mu = \left[\sum_{j=1}^{r_1+r_2} x_j + (n - r_1 - r_2)X \right] \Big/ (r_1 + r_2)$$

$$= [r_1 \bar{x}_1 + r_2 \bar{x}_2 + (n - r_1 - r_2)X]/(r_1 + r_2)$$

and

$$\hat{p} = r_1/(r_1 + r_2).$$

If the number of classes, c, is greater than 2, it is easy to extend the above maximum likelihood equations to yield

$$\hat{p}_i = \frac{r_i}{n} + \frac{\hat{k}_i(n - \Sigma r_j)}{n} \qquad i = 1, \ldots, c - 1$$

and

$$\hat{\mu}_i = \bar{x}_i + \frac{\hat{k}_i(n - \Sigma r_j)X}{r_i} \qquad i = 1, \ldots, c$$

and these can be substituted in

$$\hat{k}_i = \frac{\hat{p}_i \exp(-X/\hat{\mu}_i)}{\sum_{j=1}^{c} \hat{p}_j \exp(-X/\hat{\mu}_j)} \qquad i = 1, \ldots, c - 1$$

to yield $c - 1$ simultaneous equations which can be solved iteratively $\left(\text{recall that } 0 \leqslant \hat{k}_i \leqslant 1 \text{ and } \sum_{i=1}^{c} \hat{k}_i = 1\right)$.

Mendenhall and Hader (1958) give some results of simulation experiments for the two-class case for various values of n, p, μ_1, and μ_2. They point out that the test termination time, X, has great effect on the efficiency of the estimators – the larger X is relative to μ_1 and μ_2 the better. Needless to say, n should be as large as possible. Their final conclusion is that when n and X are small the estimates are badly biased and have large variances.

3.3 Properties of exponential mixtures

Many practically useful properties of exponential mixtures are obtained if we slightly generalize the definition of mixtures to permit negative p_i (so we are discussing general linear combinations). In particular, if $y_1, \ldots, y_n$ are independently exponentially distributed with respective parameters $\mu_1, \ldots, \mu_n$ (where we assume $\mu_i \neq \mu_j$ for all i,j) then $x = \sum_{i=1}^{n} y_i$ has the general Erlang density

$$f(x) = \sum_{i=1}^{n} \pi_i \frac{1}{\mu_i} \exp(-x/\mu_i) \tag{3.26}$$

where

$$\pi_i = \prod_{\substack{j=1 \\ j \neq i}}^{n} \mu_i/(\mu_i - \mu_j) \qquad i = 1, \ldots, n.$$

This distribution has applications in failure time and life statistics. Given that we are permitting negative p_i it is useful to have conditions indicating when such a mixture is a p.d.f. Steutel (1967) and Bartholomew (1969) have investigated this problem and have shown that (assuming, without loss of generality $\mu_1 > \mu_2 > \ldots > \mu_c$)

(i) necessary conditions for $f(x)$ to be a p.d.f. are

$$\sum_{i=1}^{c} p_i/\mu_i \geqslant 0 \text{ and } p_1 > 0;$$

(ii) sufficient conditions for $f(x)$ to be a p.d.f. are

$$\sum_{i=1}^{k} p_i/\mu_i \geqslant 0 \qquad k = 1, \ldots, c.$$

Bartholomew also obtains another set of sufficient conditions which can be written in various ways, for example:

$$\sum_{i=1}^{c} p_i/\mu_i \geqslant 0 \qquad \sum_{i=1}^{c-k} \frac{p_i}{\mu_i} \prod_{m=c-k+1}^{c} \left(\frac{1}{\mu_m} - \frac{1}{\mu_i}\right) \geqslant 0 \qquad k = 1, \ldots, c-1.$$

It should, however, be noted that both of the sets of sufficient conditions given in (ii) are, except for the case $c = 2$ [when they coincide with the conditions given in (i)] not necessary.

Behboodian (1972) presents a general theorem on the distribution of symmetric statistics from a two-component mixture:

Theorem

Let $x_1, \ldots, x_n$ be a random sample from

$$f(x) = pf_1(x) + qf_2(x)$$

and let $T = g(x_1, \ldots, x_n)$ be a symmetric statics. Let

$$T_k = g(x_{k1}, x_{k2}, \ldots, x_{kn}) \qquad k = 0, 1, \ldots, n$$

be a statistic for which the x_{ki} are independent with density $f_1(x)$ if $i \leqslant k$ and density $f_2(x)$ if $i > k$. Then

$$f_T(x) = \sum_{k=0}^{n} \binom{n}{k} p^k q^{n-k} f_{T_k}(t).$$

That is, the density of T is a binomial mixture of the densities of the T_k. □

Using this theorem, Behboodian then shows that 'the distribution of the first-order statistic for a random sample of size n from a mixture of two exponential densities with parameters μ_1 and μ_2 is a binomial mixture of $n + 1$ exponential densities with parameters

$$v_h = \left\{ \frac{k}{\mu_1} + \frac{n-k}{\mu_2} \right\}^{-1}.$$

Hill (1963) investigates the sample size required to achieve a specified level of accuracy in estimating the mixing proportion p in a two-component exponential mixture (he also deals with normal mixtures). His approach is to find expansions for the Fisher information matrix $\boldsymbol{I}$ and hence estimates of the variance of the estimates of p. He also gives a table of $\boldsymbol{I}$ values for a range of p and μ_1/μ_2 values.

In Chapter 1 we referred to Medgyessi's method for decomposing functions known to be linear combinations of components. Smith, Cohn–Sfetcu, and Buckmaster (1976) have investigated similar approaches, concentrating on linear combinations of exponential decays. Again their problem is slightly different from ours in that they begin with a function to be decomposed, without the need to estimate it from an observed distribution of measurements. They present three transformations, analogous to the Fourier transform (which identifies those frequencies contributing most to a signal), which identify the exponential decays which compose the final function. Zlokazov (1978) has also studied the decomposition of mixtures arising as spectra.

3.4 Other continuous distributions

Very little attention has been devoted to practical applications of mixtures of continuous distributions which are not normal or exponential. Nevertheless, mixtures of such distributions have very important theoretical properties and occur frequently in mathematical statistics.

3.4.1 *Non-central chi-squared distribution*

The probability density function of χ'^2, a non-central χ^2 with non-centrality parameter λ and v degrees of freedom, can be expressed as a mixture of central χ^2 densities

$$f_{\chi'^2}(x) = \sum_{j=0}^{\infty} \left[\frac{(\frac{1}{2}\lambda)^j \exp(-\frac{1}{2}\lambda)}{j!} \right] f_{\chi^2 + 2j}(x).$$

Note that the weights are the probabilities of a Poisson distribution with mean $\lambda/2$.

3.4.2 *Non-central F distribution*

The non-central F distribution is the distribution of

$$(\chi'^2_{v_1}/v_1)/(\chi^2_{v_2}/v_2)$$

where $\chi'^2_{v_1}$ is distributed as a non-central χ^2 with v_1 degrees of freedom and $\chi^2_{v_2}$ is distributed as χ^2 with v_2 degrees of freedom. Now, from the above property of non-central χ^2 distributions, we can express the non-central F in the form

$$f_{F'}(x) = f_{Fv_1,v_2}(x)\exp\left(-\tfrac{1}{2}\lambda_1\right) \sum_{j=0}^{\infty} \left\{ \left(\frac{\frac{1}{2}\lambda_1 v_1 x}{v_2 + v_1 x} \right)^j \right.$$
$$\left. \times \frac{(v_1 + v_2)(v_1 + v_2 + 2)\dots[v_1 + v_2 + 2(j-1)]}{j! v_1(v_1 + 2)\dots[v_1 + 2(j-1)]} \right\}.$$

(See, for example, Johnson and Kotz, 1970*b*, p. 191.)

The cumulative non-central F can be expressed as a mixture of incomplete beta function ratios:

$$P[F'_{v_1,v_2}(\lambda_1) \leqslant x_0] = \sum_{j=0}^{\infty} \left[\frac{(\frac{1}{2}\lambda_1)^j \exp(-\frac{1}{2}\lambda_1)}{j!} \right] I_{v_1 x_0/(v_2 + v_1 f_0)}$$
$$\left(\frac{v_1}{2} + j; \frac{v_2}{2} \right)$$

where

$$I_x(a,b) = \frac{1}{B(a,b)} \int_0^x t^{a-1}(1-t)^{b-1}\,dt.$$

The distribution of $G' = \chi^2_{\nu_1}(\lambda_1)/\chi^2_{\nu_2}$ can be expressed as a mixture of central $G_{\nu_1+2j,\nu_2} = \chi^2_{\nu_1+2j}/\chi^2_{\nu_2}$ distributions (see Johnson and Kotz):

$$f_{G'}(x) = \sum_{j=0}^{\infty} \left[\frac{\exp(-\frac{1}{2}\lambda_1)(\frac{1}{2}\lambda_1)^j}{j!} \right] f_{G_{\nu_1+2j,\nu_2}}(x).$$

Similarly, for the *doubly* non-central F (i.e. with both numerator and denominator non-central χ^2) we can express the distribution of

$$G'' = \chi'^2_{\nu_1}(\lambda_1)/\chi'^2_{\nu_2}(\lambda_2)$$

as a mixture of central G_{ν_1+2j,ν_2+2k} distributions

$$f_{G''}(x) = \sum_{j=0}^{\infty} \sum_{k=0}^{\infty} [\exp(-\tfrac{1}{2}\lambda_1)(\tfrac{1}{2}\lambda_1)^j/j!][\exp(-\tfrac{1}{2}\lambda_2)(\tfrac{1}{2}\lambda_2)^k/k!]$$
$$\times [B(\tfrac{1}{2}\nu_1 + j, \tfrac{1}{2}\nu_2 + k)]^{-1} x^{(1/2)\nu_1+j-1}(1+x)^{-(1/2)(\nu_1+\nu_2)-j-k}.$$

3.4.3 *Beta distributions*

A single beta distribution has the form

$$f(x) = \frac{\Gamma(a+b+2)}{\Gamma(a+1)\Gamma(b+1)} x^a(1-x)^b$$

with $0 < x < 1$ and $a, b > -1$.

Bremner (1978) presents an algorithm (coded in FORTRAN) for integrating mixtures of beta distributions with known parameters. The program calculates cumulative distribution functions of the forms

(i) $$F(x) = \sum_{i=0}^{n} q_i I_x\left(\frac{i}{2}, \frac{l}{2}\right) \qquad l \geqslant 1 \qquad 0 \leqslant x < 1$$

and

(ii) $$F(x) = \sum_{i=0}^{n} q_i I_x\left(\frac{i}{2}, \frac{l+n-i}{2}\right) \qquad l \geqslant 0 \qquad 0 \leqslant x < 1$$

where $I_x(a,b)$ $(a,b > 0)$ is the incomplete beta function ratio

$$I_x(a,b) = \int_0^x y^{a-1}(1-y)^{b-1}\,\mathrm{d}y/B(a,b)$$

and

$$I_x(a,0) = 0 \qquad I_x(0,b) = 1.$$

Such mixtures arise in a number of contexts, for example:

(a) We pointed out above that the cumulative distribution function of the non-central F could be expressed as an infinite mixture of form (i).

(b) Tretter and Walster (1975) show that the central cumulative density function of Wilks's Λ $(=|E|/|E+H|)$ can be written as an infinite mixture of incomplete beta functions.

(c) Bohrer and Francis (1972) develop one-sided confidence bounds for functions of the form $f(\boldsymbol{x};\boldsymbol{\beta})=\boldsymbol{x}'\boldsymbol{\beta}(x_i \geqslant 0)$ based on statistics $(\hat{\boldsymbol{\beta}}, S)$ [with $\hat{\boldsymbol{\beta}} \sim N(\boldsymbol{\beta}, \boldsymbol{B}\sigma^2)$ and $lS^2/\sigma^2 \sim \chi^2(l)$ independent of $\hat{\beta}$] which are sharper than those provided by Scheffe (1959). These functions arise, for example, in analysis of variance. Their bounds can be expressed via the coverage probability

$$P[\boldsymbol{x}'\boldsymbol{\beta} \leqslant \boldsymbol{x}'\hat{\boldsymbol{\beta}} + cS(\boldsymbol{x}'\boldsymbol{\beta}\boldsymbol{x})^{1/2}]$$

which can be rewritten as

$$\sum_{j=0}^{N} q_j I_x\left(\frac{j}{2}, \frac{l}{2}\right).$$

(d) Mixtures of type (ii) arise in calculating significance levels for the ordered $\bar{E}^2$ tests of Barlow, Bartholomew, Bremner, and Brunk (1972).

3.4.4 *Doubly non-central t distribution*

$$t''_v(\delta, \lambda) = (Z+\delta)\sqrt{v/\chi'_v(\lambda)}$$

where $Z \sim N(0,1)$ and $\chi'_v(\lambda)$ is distributed as non-central chi with v degrees of freedom and non-centrality parameter λ (see, for example, Johnson and Kotz, 1970*b*, p. 213). Then the distribution of $t''_v(\delta, \lambda)$ can be expressed as the mixture

$$\sum_{j=0}^{\infty} \left[\frac{\exp(-\frac{1}{2}\lambda)(\frac{1}{2}\lambda)^j}{j!}\right] \sqrt{[v(v+2j)^{-1}]}\, t'_{v+2j}(\delta)$$

where $t'_{v+2j}(\delta)$ is distributed as

$$(Z+\delta)\sqrt{v/\chi_v}.$$

3.4.5 *Planck's distribution*

This distribution arises in some areas of physics. Its standardized form is

$$f(x) = C_k x^k/(\mathrm{e}^x - 1) \qquad 0 < x$$

which can be rewritten as

$$f(x) = C_k x^k \sum_{j=1}^{\infty} e^{-jx}.$$

Thus the Planck distribution is a mixture of distributions of $\chi^2_{2(k+1)}/2j$ with weights proportional to j^{-k-1} $(j = 1, 2, \ldots)$ (Johnson and Kotz, 1970*b*, p.274).

3.4.6 *Logistic*

Shah (1963) uses the method of moments to estimate the parameters of a two-component logistic mixture with results very similar to those of Rider (1961).

3.4.7 *Laplace*

This distribution is also called the double exponential (and for other names see Johnson and Kotz, 1970*b*, p. 22). Krysicki (1966) considers two-component mixtures, giving formulae for estimating μ_1 and μ_2.

3.4.8 *Weibull*

Kao (1959) uses a two-component Weibull mixture to fit the life distribution of electron valves (cf. the earlier discussion of exponential mixtures). He bases his analysis on the observation that there are two types of failure, sudden catastrophic failures and wear-out failures, with respective distribution functions

$$G_1(x) = 1 - \exp(-x^{\beta_1}/\alpha_1)$$

(with $x > 0, \alpha_1 > 0, 0 < \beta_1 < 1$) and

$$G_2(x) = 1 - \exp[-(x - \gamma)^{\beta_2}/\alpha_2]$$

(with $x > \gamma, \alpha_2 > 0, \beta_2 > 1$). He analyses this mixture using a graphical approach. Since we believe that readers will find it more convenient to use one of the general methods already outlined we shall not give his method in its entirety, but content ourselves with estimating the mixing proportion p_1. This is done as follows:

(a) Plot the sample cumulative per cent failed function of the data on ln–ln versus ln paper and by hand fit a smooth curve to the points.

(b) Draw lines tangent to each end of this curve (these tangents estimate $p_1 G_1$ and $p_2 G_2$).

(c) At the intersection of the tangent estimating $p_2 G_2$ with the upper borderline (i.e. approximately where the fitted curve cuts the upper borderline) drop a vertical line to intersect the other tangent. The co-ordinate of this intersection on the per cent failure scale gives $\hat{p}_1$, the estimate of p_1.

Falls (1970) proposes the mixed Weibull distribution as a model for atmospheric data and applies the method of moments to the two-component case:

$$f(x) = p_1 g_1(x) + p_2 g_2(x)$$

where

$$f_i(x) = \frac{\beta_i}{\alpha_i} x^{\beta_i - 1} \exp\left(-\frac{x^{\beta_i}}{\alpha_i} \right).$$

The first five theoretical moments about the origin take the form

$$\mu_r = p_1 \alpha_1^{r/\beta_1} \Gamma\left(\frac{r}{\beta_1} + 1 \right) + p_2 \alpha_2^{r/\beta_2} \Gamma\left(\frac{r}{\beta_2} + 1 \right) \qquad r = 1, \ldots, 5$$

where Γ is the gamma function. In the usual way, Falls equates these to the sample moments m_r' and then, by the substitutions

$$v = \alpha_1^{1/\beta_1}$$

$$w = \alpha_2^{1/\beta_2}$$

$$\theta_r = \Gamma\left(\frac{r}{\beta_1} + 1 \right) \qquad r = 1, 2, 3$$

$$\Psi_r = \Gamma\left(\frac{r}{\beta_2} + 1 \right) \qquad r = 1, 2, 3$$

reduces the first three moment equations to

$$m_r' = p_1 v^r \theta_r + p_2 w^r \Psi_r \qquad r = 1, 2, 3.$$

Solving the first of these for v and substituting in the second gives a quadratic in w which can be solved and substituted back into the first to give w and v. Finally, substituting these into the third equation gives a single equation for the three unknowns p_1, β_1, and β_2. Falls then suggests estimating p_1 by Kao's graphical method, outlined above, and estimating β_1 and β_2 using the Newton–Raphson technique. As with other authors, he adopts the suggestion of choosing between multiple solutions by comparing higher-order moments.

In Chapter 1 we commented about the difficulties associated with using high-order moments. In his conclusion Falls gives an illustration of this. He says 'a comparison of sample moments for the sample showed an error of 0.09 per cent for the first moment, 0.44 per cent for the second moment, 1.64 per cent for the third moment, 4.94 per cent for the fourth moment, and 12.73 per cent for the fifth moment.'

Padgett and Tsokos (1978) and Harris and Singpurwalla (1968) also consider Weibull mixtures, but since they use continuous functions as mixing distributions we shall not discuss them.

3.4.9 *Gamma*

John (1970) briefly outlines the use of the methods of moments and maximum likelihood in estimating the parameters of two-component gamma mixtures.

3.5 Mixtures of different component types

There seem to be few applications involving mixtures where the components are of different types. One reason for this is presumably that there must be strong *a priori* theoretical grounds for considering such a mixture since mixtures of flexible components of the same type (such as Weibull) can usually be found which will provide an adequate fit to a given data set.

However, Ashton (1971) applies such a mixture in studying the distribution of time gap in road traffic flow. Her model is as follows. She assumes that behind each vehicle is a 'zone of emptiness' of length Z which vehicles never enter. A proportion p of vehicles stay at exactly distance Z from the preceding vehicle. The remainder (proportion $1-p$) follow at a distance Y behind the preceding vehicle, where Y is a displaced exponential distribution

$$f_1(Y)=\frac{1}{\mu}\exp[-(Y-Z)/\mu].$$

Under the assumption that Z follows a gamma distribution

$$f_2(Z)=\frac{1}{\beta^K\Gamma(K)}Z^{K-1}\exp(-Z/\beta) \qquad Z>0$$

the overall p.d.f. of the time gap is

$$f(x) = \frac{px^{K-1}\exp(-x/\beta)}{\beta^K\Gamma(K)} + (1-p)\frac{\exp(-x/\mu)\gamma(K,x/\beta)(1+\beta/\mu)^K}{\mu\Gamma(K)}$$

$$x \geqslant 0.$$

If the gamma distribution is replaced by a Pearson type III distribution in order to remove the possibility of zero gaps, we have

$$f(x) = \frac{p(x-\alpha)^{K-1}\exp[-(x-\alpha)/\beta]}{\beta^K\Gamma(K)}$$

$$+(1-p)\frac{\exp[-(x-\alpha)/\mu]\gamma\left(K,\frac{x-\alpha}{\beta}\right)(1+\beta/\mu)^K}{\mu\Gamma(K)}$$

$$x \geqslant \alpha.$$

3.6 Summary

Most of the work on mixtures with continuous components which are not normal has been on exponential components. Such mixtures arise in industrial applications, notably in the analysis of failure time data, and have important mathematical properties.

The standard approaches of moments and maximum likelihood have been applied in estimating the mixture parameters. Several other methods which have formal similarities to the moments method have also been devised – in particular, one which makes use of fractional moments and appears to possess much greater efficiency than the usual moments method. Of course, the difficulties of using the method of moments to estimate parameters of mixtures of many components, outlined in Chapter 1, still apply.

Mixtures of other forms of continuous component seem rare as far as practical applications go, but such mixtures have importance in theoretical problems. Practical applications of mixtures of different types of components also seem rare in the literature.

CHAPTER 4

Mixtures of discrete distributions

4.1 Introduction

In this chapter, mixtures of certain *discrete* distributions will be considered. As with mixtures of normals, Pearson (1915) appears to have been the first to study such distributions in any detail, deriving moment estimators for the parameters in a mixture of binomial density functions. A large part of this chapter will involve mixtures in which the components are binomial or Poisson distributions, since these have been extensively studied in the literature. However, mixtures of other discrete distributions will be considered briefly, in particular the multivariate Bernoulli distribution which, as noted by Wolfe (1969), may be used as the basis of latent class analysis (see Lazarsfeld, 1968). We begin by considering mixtures of binomial distributions.

4.2 Mixtures of binomial distributions

Let $X_1, \ldots, X_n$ be independent and identically distributed random variables each having probability density

$$P(x\,;\boldsymbol{p},\boldsymbol{\theta}) = \text{prob}(X_i = x) = \sum_{j=1}^{c} p_j P_j(x\,;\theta_j) \tag{4.1}$$

where

$$P_j(x\,;\theta_j) = \binom{m}{x}\theta_j^x(1-\theta_j)^{m-x} \tag{4.2}$$

$(x = 0, 1 \ldots, m)$, with $0 < \theta_1 < \theta_2 \ldots < \theta_c < 1, 0 < p_j < 1 (j = 1, \ldots, c)$ and $\sum_{j=1}^{c} p_j = 1$. (The assumption that we are able to order the parameters θ_j simply implies that these are distinct.)

The probability density function given in (4.1) is a mixture of c binomial distributions with $2c - 1$ parameters, $\boldsymbol{\theta} = (\theta_1, \ldots, \theta_c)$ and

$\boldsymbol{p} = (p_1, \ldots, p_{c-1})$, and the first problem is to construct estimators for these. However, before estimation can be attempted it must be verified that the mixture is *identifiable*, that is it must be uniquely characterized so that two distinct sets of parameters cannot yield the same mixed distribution. This problem has been considered for mixtures in general in Chapter 1, and it will suffice to mention here that a necessary and sufficient condition that the mixture defined by (4.1) be identifiable is that $m \geqslant 2c - 1$. The necessity of this condition was established by Teicher (1961) and its sufficiency by Blischke (1964). Consequently in the following accounts of estimation procedures we shall assume that the condition holds.

4.2.1 *Moment estimators for binomial mixtures*

Much of the work on binomial mixtures reported in the literature has involved the derivation of moment estimators for the parameters. The rationale for concentrating on this method rather than the more statistically respectable maximum likelihood approach has been that the latter method yields 'highly intractable equations' or some similar argument. In these days of very powerful computers such comments are, of course, no longer really justified, and maximum likelihood methods may be used with no great difficulty in many cases, and they will indeed be discussed in detail in Section 4.2.2. However, it remains of interest to consider the estimators given by the method of moments, since they may be extremely useful in providing initial parameter values for the iterative algorithms involved in deriving maximum likelihood estimates.

The moment estimators for mixtures of binomial distributions are most easily obtained by considering the first $2c - 1$ sample 'factorial moments' of (4.1) defined as follows:

$$V_k = \frac{1}{n} \sum_{i=1}^{n} \frac{X_i(X_i - 1)(X_i - k + 1)}{m(m-1)\ldots(m-k+1)} \tag{4.3}$$

$$k = 1, \ldots, 2c - 1.$$

(These differ by a constant multiplier from the quantities usually defined to be factorial moments; see Section 4.3.)

It is easy to show that the corresponding population moments, v_k, are given by

$$v_k = E(V_k) = \sum_{j=1}^{c} p_j \theta_j^k. \tag{4.4}$$

To construct the moment estimators the observed values $V_1, \ldots, V_{2c-1}$ are equated to the expected values $v_1, \ldots, v_{2c-1}$, and the resulting equations solved for $p_1, \ldots, p_c$ and $\theta_1, \ldots, \theta_c$, subject to the conditions that $\sum_{j=1}^{c} p_j = 1$. The details of the procedure are as follows.

Equating observed and expected factorial moments we have

$$V_k = \sum_{j=1}^{c} \hat{p}_j \hat{\theta}_j^k \qquad k = 0, \ldots, 2c-1 \tag{4.5}$$

where V_0 is defined to be unity and $\hat{p}_j$ and $\hat{\theta}_j$ are estimates of the corresponding parameters in (4.1). These equations are linear in the mixing proportions, p_i, and therefore these may be eliminated by solving algebraically any $c - 1$ of the equations for them, and substituting the solution in the remaining c equations. (This procedure has been previously described in Chapter 1 and used in Chapters 2 and 3.)

This results in a series of c simultaneous polynomials in the c unknowns, $\hat{\theta}_1, \ldots, \hat{\theta}_c$. Blischke (1964) shows that this system of equations may be reduced to a single polynomial of degree c:

$$x^c + B_{c-1} x^{c-1} \ldots + B_1 x + B_0 = 0 \tag{4.6}$$

the coefficients of which are given by the solution of the equations

$$\boldsymbol{Z\beta} = \boldsymbol{\phi} \tag{4.7}$$

where

$$\boldsymbol{Z} = \begin{bmatrix} 1 & V_1 \ldots V_{c-1} \\ V_1 & \\ \vdots & \\ V_{c-1} & V_{2c-2} \end{bmatrix} \qquad \boldsymbol{\beta} = \begin{bmatrix} B_0 \\ B_1 \\ \vdots \\ B_{c-1} \end{bmatrix} \qquad \boldsymbol{\phi} = \begin{bmatrix} -V_c \\ -V_{c+1} \\ \vdots \\ -V_{2c-1} \end{bmatrix}$$

Blischke shows that Z is non-singular so that (4.7) may be solved to give the coefficients of (4.6), which is then itself solved by one of the methods mentioned in Chapter 2, to give the parameter estimates $\hat{\theta}_1, \ldots, \hat{\theta}_c$. (It should be mentioned that there is no guarantee that Equation 4.6 will have c roots in the interval (0, 1), so that the estimates $\hat{\theta}_1, \ldots, \hat{\theta}_c$ may be negative, imaginary, or greater than one. However, the probability of this tends to zero as the sample size, n, tends to infinity).

Having determined the estimates $\hat{\theta}_1, \ldots, \hat{\theta}_c$, estimates of the mixing proportions may be obtained by solving the first c equations of (4.5). That is, we solve the equations

$$\boldsymbol{\Theta}\hat{\boldsymbol{p}} = \boldsymbol{\delta} \tag{4.8}$$

for $\hat{\boldsymbol{p}}$, where

$$\boldsymbol{\Theta} = \begin{bmatrix} 1 & 1 & 1 \\ \hat{\theta}_1 & \hat{\theta}_2 & \ldots \hat{\theta}_c \\ \hat{\theta}_1^2 & \hat{\theta}_2^2 & \ldots \hat{\theta}_c^2 \\ & & \\ \hat{\theta}_1^{c-1} & \hat{\theta}_2^{c-1} & \ldots \hat{\theta}_c^{c-1} \end{bmatrix} \quad \hat{\boldsymbol{p}} = \begin{bmatrix} \hat{p}_1 \\ \hat{p}_2 \\ \\ \\ \hat{p}_c \end{bmatrix} \quad \boldsymbol{\delta} = \begin{bmatrix} 1 \\ V_1 \\ V_2 \\ \vdots \\ V_{c-1} \end{bmatrix}.$$

Again there is no guarantee that each of the mixing proportions will lie in the interval (0, 1).

In the case of mixtures of *two* binomial distributions, simple explicit formulae may be obtained for the estimates of p_1, θ_1, and θ_2 in terms of the first three moments. These are as follows:

$$\hat{p}_1 = (V_1 - \hat{\theta}_1)/(\hat{\theta}_1 - \hat{\theta}_2) \tag{4.9}$$

$$\hat{\theta}_1 = \tfrac{1}{2}A - \tfrac{1}{2}(A^2 - 4AV_1 + 4V_2)^{1/2} \tag{4.10}$$

$$\hat{\theta}_2 = \tfrac{1}{2}A + \tfrac{1}{2}(A^2 - 4AV_1 + 4V_2)^{1/2} \tag{4.11}$$

where

$$A = (V_3 - V_1 V_2)/(V_2 - V_1^2).$$

Since these estimates can assume complex as well as indeterminate values, Blischke suggests using them only when $A^2 - 4AV_1 + 4V_2 > 0$ and $(A^2 - 4AV_1 + 4V_2)^{1/2} \leqslant \min(A, 2 - A)$. In other cases he suggests that only a *single* binomial distribution should be fitted with, of course, $p_2 = 0$ and $\hat{\theta}_1 = V_1$.

Example

To illustrate this estimation procedure, let us consider the data shown in Table 4.1, which is generated from a mixture of four binomial distributions with the following parameter values:

$$\begin{array}{llll} p_1 = 0.2 & p_2 = 0.2 & p_3 = 0.2 & \\ \theta_1 = 0.1 & \theta_2 = 0.2 & \theta_3 = 0.6 & \theta_4 = 0.9. \end{array}$$

m has the value 30 and n the value 200. The first seven 'factorial

Table 4.1. 200 *variates generated from a mixture of four binomial distributions.*

7	25	27	27	24	25	12	6	29	5
7	29	6	2	24	15	15	27	4	5
26	28	5	16	29	7	28	28	27	29
10	28	29	30	1	30	5	28	7	2
3	28	28	22	21	30	27	4	6	25
2	28	16	7	2	2	17	3	3	3
29	27	4	3	1	23	23	28	28	27
25	18	21	28	26	22	28	16	26	29
19	20	24	17	3	17	26	18	6	28
28	28	29	6	27	4	27	20	28	5
30	29	29	27	26	3	10	28	22	7
29	28	24	28	12	22	2	27	5	28
28	14	13	27	22	3	7	7	28	3
8	2	27	5	4	29	29	2	1	3
6	1	3	6	21	17	6	10	29	4
1	25	0	7	28	29	28	28	29	1
28	28	26	28	24	8	19	28	25	3
17	19	10	0	9	15	15	26	20	26
6	27	22	2	27	14	14	4	9	28
27	28	28	3	28	28	27	24	27	27

moments' take the values

$$V_1 = 0.5890, \quad V_2 = 0.4650, \quad V_3 = 0.3981 \quad V_4 = 0.3502$$
$$V_5 = 0.3122, \quad V_6 = 0.2803 \quad V_7 = 0.2528.$$

and from these the following parameter estimates are obtained:

$$\hat{p}_1 = 0.0959 \quad \hat{p}_2 = 0.2488 \quad \hat{p}_3 = 0.1687$$
$$\hat{\theta}_1 = 0.0670 \quad \hat{\theta}_2 = 0.1777 \quad \hat{\theta}_3 = 0.5694 \quad \hat{\theta}_4 = 0.9090$$

These will be compared with the maximum likelihood estimates for the same data set in Section 4.3. Here we move on to consider briefly the asymptotic properties of the moment estimators. □

Asymptotic properties of the moment estimators

Blischke (1962, 1964) showed that the moment estimators for mixtures of binomial distributions are asymptotically normally distributed with means equal to the population parameter values and covariance matrix given by

$$\boldsymbol{\Sigma} = \boldsymbol{\Psi}_c \boldsymbol{\Sigma}_V^{(2c-1)} \boldsymbol{\Psi}'_c \tag{4.12}$$

where

$$\mathbf{\Psi}_c^{-1} = \begin{bmatrix} p_1 & \dots p_c & \theta_1 - \theta_c & \dots \theta_{c-1} - \theta_c \\ 2p_1\theta_1 & 2p_c\theta_c & \theta_1^2 - \theta_c^2 & \theta_{c-1}^2 - \theta_c^2 \\ \vdots & & & \\ (2c-1)p_1\theta_1^{2c-2} & \dots (2c-1)p_1\theta_c^{2c-2} & \theta_1^{2c-1} - \theta_c^{2c-1} & \dots \theta_{c-1}^{2c-1} - \theta_c^{2c-1} \end{bmatrix}$$

and $\mathbf{\Sigma}_V^{(2c-1)}$ is the covariance matrix of the factorial moments given by Equation (4.3); the elements of this matrix may be obtained directly since

$$\operatorname{var}(V_k) = [nm^{(k)}]^{-1}(m-k)^{(k-1)}\left[(m-2k+1)\sum_{j=1}^{c} p_j\theta_j^{2k} + k^2\sum_{j=1}^{c} p_j\theta_j^{2k-1}\right] - \frac{1}{n}\left[\sum_{j=1}^{c} p_j\theta_j^{k}\right]^2 + O(m^{-2}) \tag{4.13}$$

$$\operatorname{cov}(V_k, V_{k'}) = [nm^{(k)}m^{(k')}]^{-1}n^{(k+k'-1)}[(n-k-k'+1)\textstyle\sum p_j\theta_j^{k+k'} + kk'\sum p_j\theta_j^{k+k'-1}] - \frac{1}{n}[\sum p_j\theta_j^k][\sum p_j\theta_j^{k'}] + O(m^{-2}). \tag{4.14}$$

Blischke (1964) shows that these expressions can be combined to yield the (k, k') entry of $\mathbf{\Sigma}_V^{(2c-1)}$ in the form

$$\mu_{11}(k, k') = \frac{kk'}{n}(v_{k+k'-1} - v_{k+k'}) + (v_{k+k'} - v_k v_{k'}) + O(m^{-2}). \tag{4.15}$$

Explicit expressions for $\mathbf{\Sigma}$ in the case $c = 2$ are given in Blischke (1962).

Blischke also considers the joint asymptotic efficiency of the asymptotically normally distributed moment estimators relative to the corresponding maximum likelihood estimators. This is found simply as the ratio of the determinants of the covariance matrices of the two sets of estimators. Blischke finds that the asymptotic efficiencies as functions of the binomial parameter m exhibit a rather unusual behaviour. The efficiencies are unity if m takes its lowest possible value, $(2c - 1)$, less than unity for m greater than its minimum value, but again tend to unity as m becomes large. (These asymptotic results refer to limits as n tends to infinity.)

An estimate of $\mathbf{\Sigma}$ is of course provided by substituting parameter estimates for population values in Ψ_c and $\mathbf{\Sigma}_V^{(2c-1)}$.

4.2.2 *Maximum likelihood estimators for mixtures of binomial distributions*

The maximum likelihood procedure for the general finite mixture distribution when applied to mixtures of binomial distributions leads to the following estimation equations:

$$\hat{p}_r = \frac{1}{n} \sum_{i=1}^{n} \hat{P}(r|X_i) \tag{4.16}$$

$$\hat{\theta}_r = \frac{1}{nm\hat{p}_r} \sum_{i=1}^{n} X_i \hat{P}(r|X_i) \tag{4.17}$$

where

$$\hat{P}(r|X_i) = \frac{\hat{p}_r P_r(x;\hat{\theta}_r)}{P(x, \hat{\boldsymbol{p}}, \hat{\boldsymbol{\theta}})}. \tag{4.18}$$

Given initial values for $\hat{\boldsymbol{p}}$ and $\hat{\boldsymbol{\theta}}$, Equations (4.16) and (4.17) may be used as the basis of an iterative estimation algorithm analogous to that described in previous chapters.

Example

Applying such a procedure to the data shown in Table 4.1, using as initial parameter values first the population values and second the estimates given by the method of moments, leads to the final parameter estimates shown in Table 4.2.

Table 4.2. *Maximum likelihood estimates for the data shown in Table* 4.1.

Starting values	*Number of iterations*	*Final values*
(1) $p_1 = 0.2$	6	0.181 78
$p_2 = 0.2$		0.179 12
$p_3 = 0.2$		0.175 37
$\theta_1 = 0.1$		0.090 49
$\theta_2 = 0.2$		0.218 84
$\theta_3 = 0.6$		0.604 56
$\theta_4 = 0.9$		0.912 17
(2) $p_1 = 0.095\,10$	27	0.183 15
$p_2 = 0.248\,84$		0.178 00
$p_3 = 0.168\,71$		0.175 55
$\theta_1 = 0.066\,97$		0.090 88
$\theta_2 = 0.177\,73$		0.219 68
$\theta_3 = 0.569\,41$		0.605 18
$\theta_4 = 0.909\,04$		0.912 32

Here the two sets of final estimates are very similar. This will not always be the case, and with less clearly separated component distributions problems may arise because different initial values may lead to widely different final estimates. The value of the likelihood function must then be used to indicate which of these is to be preferred. □

The asymptotic covariance matrix of the maximum likelihood estimates may be obtained as usual from the inverse of the information matrix; Blischke (1962, 1964) shows that as the binomial term, m, tends to infinity, this matrix tends to

$$\left[\begin{array}{ccc|cccc}
\frac{mp_1}{\theta_1(1-\theta_1)} & & & & & & \\
 & \mathbf{O} & & & & & \\
 & \frac{mp_2}{\theta_2(1-\theta_2)} & & & & \mathbf{O} & \\
 & & \ddots & & & & \\
\mathbf{O} & & \frac{mp_c}{\theta_c(1-\theta_c)} & & & & \\
\hline
 & & & \frac{1}{p_1}+\frac{1}{p_c} & \frac{1}{p_c} & \cdots & \frac{1}{p_c} \\
 & & & \frac{1}{p_c} & \frac{1}{p_2}+\frac{1}{p_c} & \cdots & \frac{1}{p_c} \\
 & \mathbf{O} & & \vdots & & & \\
 & & & \frac{1}{p_c} & \frac{1}{p_c} & \frac{1}{p_{c-1}} & +\frac{1}{p_c}
\end{array}\right]. \tag{4.19}$$

In a small simulation investigation, Hasselblad (1969) found that sample variances of the maximum likelihood estimates are uniformly smaller than sample variances of the moment estimates. A further advantage with the iterative maximum likelihood procedure is that it is impossible to obtain estimates outside the admissible limits for the parameters, if the initial values are chosen to fall within these limits.

4.2.3 *Other estimation methods for mixtures of binomial distributions*

Blischke (1964) considers a number of alternatives to the method of moments for estimating the parameters in a mixture of binomial distributions. The first of these is based upon a $\min(\chi^2)$ approach but avoids having to minimize the χ^2 statistic by equating its derivatives with respect to the parameters to zero; instead χ^2 is expanded in a Taylor's series about some guessed solution, and truncated after the quadratic term, this leading to an equation which may be solved to give the necessary adjustment to the original estimate. Blischke shows that by using the moment estimate as the initial solution, a single iteration of the process leads to an asymptotically efficient estimate.

A further procedure suggested by Blischke is related to the method of scoring (see Rao, 1965), sometimes used to solve maximum likelihood equations. In this case, however, only one iteration of the process is performed, again using the moment estimate to provide the initial value. Again this will lead to an asymptotically efficient estimator. A small empirical study by Blischke shows that the estimates produced by this scheme are, in general, more satisfactory than those given by the modified minimum χ^2 approach and by the method of moments.

These methods are now probably only of historical interest, since with modern computers the maximum likelihood estimates may be obtained without great difficulty, using perhaps the moment estimates as initial values.

4.3 Mixtures of Poisson distributions

Let $X_1, X_2, \ldots, X_n$ be independent and identically distributed random variables each having probability density

$$P(x\,;\mathbf{p},\boldsymbol{\lambda}) = \text{prob}(X_i = x) = \sum_{j=1}^{c} p_j P_j(x\,;\lambda_j) \ldots \qquad (4.20)$$

where

$$P_j(x\,;\lambda_j) = \frac{\lambda_j^x \exp(-\lambda_j)}{x\,!} \qquad (4.21)$$

$(x = 0, 1, \ldots)$, with $0 < \lambda_1 < \lambda_2 \ldots < 1, 0 < p_j < 1 (j = 1, \ldots, c)$ and $\sum_{j=1}^{c} p_j = 1$.

The probability density function given in (4.20) is a mixture of c Poisson distributions with $2c-1$ parameters $\boldsymbol{\lambda} = (\lambda_1, \lambda_2, \ldots, \lambda_c)$ and $\boldsymbol{p} = (p_1, p_2, \ldots, p_{c-1})$, and again we wish to construct estimators for the parameters. This is possible since the identifiability of finite mixtures of Poisson distributions has been established by Teicher (1960) and Feller (1943). We begin as previously with a discussion of moment estimators.

4.3.1 *Moment estimators for mixtures of Poisson distributions*

Several authors, for example, Rider (1961) and Cohen (1963, 1965) have considered mixtures of two Poisson distributions and derived fairly simple formulae for parameter estimation by equating either ordinary or factorial moments to expected values. However, there appears to be little discussion in the literature on Poisson mixtures with $c > 2$, although the estimation procedure follows exactly that for the binomial distribution, but with a slightly different form for the factorial moments used. If in this case we introduce the moments, V_k, given by

$$V_k = \frac{1}{n} \sum_{i=1}^{n} X_i(X_i - 1) \ldots (X_i - k + 1) \ldots \tag{4.22}$$

with expectation v_k, where

$$v_k = E(V_k) = \sum_{j=1}^{c} p_j \lambda_j^k \tag{4.23}$$

then the population value v_k is seen to be of the same form as (4.4) and, consequently, the process of equating the values $V_1, V_2, \ldots, V_{2c-1}$, to $v_1, v_2, \ldots, v_{2c-1}$ and solving for $\boldsymbol{\lambda}$ and $\boldsymbol{p}$ leads to exactly the same estimation procedure as before. In the case of $c = 2$, Equations (4.9) to (4.11) hold with $\hat{\lambda}_i$ replacing $\hat{\theta}_i$.

Example

To illustrate how the procedure works in practice let us consider the data shown in Table 4.3 which is taken from Hasselblad (1969) and gives the numbers of death notices of women 80 years of age and over, appearing in *The Times*, London, on each day for three consecutive years. A mixture of two Poissons might be considered here, since it is likely that the death rate during winter months will be higher than during the summer. Fitting a single Poisson distribution to

Table 4.3. *Number of death notices for women over* 80 *years of age appearing in* The Times, *and the results of fitting a single and a mixture of two Poisson distributions.*

Observed death count	*Observed frequency*	*Single Poisson Expected frequency*	*Mixture Expected frequency*
0	162	126.79	163.62
1	267	273.47	267.78
2	271	294.92	260.46
3	185	212.04	192.84
4	111	114.34	115.83
5	61	49.32	57.91
6	27	17.73	24.57
7	8	5.46	9.00
8	3	1.47	2.89
9	1	0.35	0.83
		$\chi_6^2 =$ 26.97	$\chi_4^2 =$ 1.52

these data gives the expected frequencies shown in Table 4.3 with a chi-square value of 26.97 with six degrees of freedom, indicating a rather poor fit. Fitting a mixture of two Poissons by the procedure outlined above leads to parameter estimates

$$p_1 = 0.287\,05 \qquad \lambda_1 = 1.102\,09 \qquad \lambda_2 = 2.581\,64$$

and to the expected frequencies also shown in Table 4.3. The chi-square value is now 1.52 with four degrees of freedom, indicating a greatly improved fit. (In the calculation of the two chi-square values the last three categories have been amalgamated.) □

Example

As a further example, moment estimates were found for the data shown in Table 4.4, generated from a three-component Poisson mixture with parameter values

$$p_1 = 0.3 \qquad p_2 = 0.3$$
$$\lambda_1 = 0.5 \qquad \lambda_2 = 3.0 \qquad \lambda_3 = 6.0.$$

The parameter estimates found were

$$\hat{p}_1 = 0.158 \qquad \hat{p}_2 = 0.604$$
$$\hat{\lambda}_1 = 0.143 \qquad \hat{\lambda}_2 = 2.921 \qquad \hat{\lambda}_3 = 7.165.$$

Table 4.4. *Data generated from a three-component Poisson mixture.*

x	*Frequency*	x	*Frequency*
0	77	9	12
1	81	10	9
2	57	11	5
3	64	12	4
4	71	13	1
5	38	14	2
6	33	15	1
7	29		
8	16		500

The rather large discrepancy between the parameter estimates and their population values may be due to the sample size, and indicates that for accurate estimates very large samples will in general be necessary. □

Rider (1962) considers the asymptotic variances of the moment estimators for mixtures of two Poisson distributions when the mixing proportion is known, and derives the expression

$$V(\hat{\lambda}_1) = [4np_1^2(\lambda_1 - \lambda_2)]^{-1}[2(p_1\lambda_1^2 + p_2\lambda_2^2) + 4p_1\lambda_1(\lambda_1 - \lambda_2)^2 - p_1p_2(\lambda_1 - \lambda_2)^4]. \tag{4.24}$$

(The variance of λ_2 may be obtained simply by interchanging p_1 and p_2 and λ_1 and λ_2 in the above expression.)

4.3.2 *Maximum likelihood estimators for a Poisson mixture*

The maximum likelihood equations in this case take the form

$$\hat{p}_r = \frac{1}{n}\sum_{i=1}^{n} \hat{P}(r|X_i) \tag{4.25}$$

$$\hat{\lambda}_r = \frac{1}{n\hat{p}_r}\sum_{i=1}^{n} X_i\hat{P}(r|X_i) \tag{4.26}$$

where

$$\hat{P}(r|X_i) = \frac{\hat{p}_r P_r(x, \hat{\lambda}_r)}{P(x, \hat{\mathbf{p}}, \hat{\boldsymbol{\lambda}})}. \tag{4.27}$$

Example
The iterative algorithm previously discussed may again be used to derive parameter estimates, and we shall illustrate its use with data consisting of 200 variates generated from a four-component Poisson mixture with parameter values

$$\begin{array}{llll} p_1 = 0.2 & p_2 = 0.2 & p_3 = 0.2 & \\ \lambda_1 = 0.5 & \lambda_2 = 1.5 & \lambda_3 = 3.0 & \lambda_4 = 5.0. \end{array}$$

The parameter estimates resulting from two sets of initial values are shown in Table 4.5. The difference in the two sets of final values is, in

Table 4.5. *Maximum likelihood estimates for a four-component Poisson mixture.*

Starting values	*Number of iterations*	*Final values*
(1) $p_1 = 0.20$	192	0.15
$p_2 = 0.20$		0.19
$p_3 = 0.20$		0.20
$\lambda_1 = 0.50$		0.75
$\lambda_2 = 1.50$		1.34
$\lambda_3 = 3.00$		3.37
$\lambda_4 = 5.00$		5.08
(2) $p_1 = 0.30$	365	0.07
$p_2 = 0.30$		0.36
$p_3 = 0.10$		0.19
$\lambda_1 = 1.00$		0.02
$\lambda_2 = 2.00$		1.31
$\lambda_3 = 4.50$		4.10
$\lambda_4 = 6.00$		4.87

this case, fairly considerable, and one would need to judge between them using the two likelihood values. It is also notable that for these data the rate of convergence of the iterative algorithm is extremely slow. A further point of interest is that the previously described moment estimation procedure fails to give a solution for these data since the estimation polynomial equivalent to (4.6) for Poisson mixtures has complex roots. This failure probably arises from the poor sampling properties of the higher moments, and illustrates the point made in the previous section as to one of the advantages of the maximum likelihood procedure. □

4.4 Mixtures of Poisson and binomial distributions

Cohen (1963) has considered a mixture distribution formed from a Poisson component and a binomial component. The probability density function of the mixture may be written as

$$P(x) = \frac{p\lambda^x \exp(-\lambda)}{x!} + (1-p)\binom{n}{x}\theta^x(1-\theta)^{n-x}. \tag{4.28}$$

In this case the expected value of the kth factorial moment given by (4.22) is as follows:

$$v_k = E(V_k) = p\lambda^k + (1-p)n(n-1)\dots(n-k+1)\theta^k. \tag{4.29}$$

Equating population and sample values in the usual way leads to the estimation equations

$$\left.\begin{aligned} V_1 - n\theta &= p(\lambda - n\theta) \\ V_2 - n(n-1)\theta^2 &= p[\lambda^2 - n(n-1)\theta^2] \\ V_3 - n(n-1)(n-2)\theta^3 &= p[\lambda^3 - n(n-1)(n-2)\theta^3]. \end{aligned}\right\} \tag{4.30}$$

On eliminating p, the mixing proportion, between the first and second and between the first and third equations of (4.30), we obtain the following simultaneous equations for λ and θ:

$$\left.\begin{aligned} V_2(\lambda - n\theta) - n\theta\lambda[(n-1)\theta - \lambda] &= V_1[\lambda^2 - n(n-1)\theta^2] \\ V_3(\lambda - n\theta) - n\theta\lambda[(n-1)(n-2)\theta^2 - \lambda^2] &= V_1[\lambda^3 - n(n-1)(n-2)\theta^3] \end{aligned}\right\}. \tag{4.31}$$

These two equations will give estimates of λ and θ, and the estimate of p then follows from the first equation of (4.30).

There do not appear to have been any attempts to generalize (4.28) to a mixture involving more than a single Poisson and binomial component.

4.5 Mixtures of other discrete distributions

A number of other mixed distributions have been considered in the literature. For example, Cohen (1963) considers a mixed truncated Poisson distribution with missing zero classes. This has the form

$$P(x) = \frac{p\lambda_1^x \exp(-\lambda_1^x)}{x![1-\exp(-\lambda_1)]} + \frac{(1-p)\lambda_2^x \exp(-\lambda_2^x)}{x![1-\exp(-\lambda_2)]}. \quad (4.32)$$

The method of moments may again be used to find parameter estimates. Rider (1961) considers a distribution formed by mixing two negative binomials, and derives estimators for the parameters, again using the method of moments. Blischke (1964) considers a mixture of binomial distributions where the mixing parameters are known. He shows that the moment estimators in this case have the strange property that their asymptotic efficiencies tend to zero rather than one as $n \to \infty$.

Cohen (1966) considers a mixture distribution of the form

$$P(x\,;p,\boldsymbol{\lambda}) = \begin{cases} (1-p) + pP_1(0\,;\boldsymbol{\lambda}) & x=0 \\ pP_1(x\,;\boldsymbol{\lambda}) & x=1,2,\ldots \end{cases} \quad (4.33)$$

where $P_1(x\,;\lambda)$ is any discrete distribution defined over the domain $x = 0, 1, 2, \ldots$ with parameters $\boldsymbol{\lambda} = (\lambda_1, \lambda_2, \ldots, \lambda_k)$. Cohen considers the special case where P_1 is a negative binomial distribution, and in an earlier paper (Cohen, 1960) where P_1 is a Poisson distribution. Estimates of the parameters may be obtained simply by maximum likelihood. This particular mixture distribution is useful when, for example, the zero class of a data set is inflated by the inclusion of individuals belonging to a 'non-infected' or 'non-susceptible' group.

Dawid and Skene (1979) consider a mixture of multinomial distributions arising in a model of observer rating, and use the EM algorithm described previously to estimate parameters.

An interesting suggestion, seemingly first made by Wolfe (1969), is that latent class analysis (see, for example, Green, 1951, Gibson, 1959, and Lazarsfeld, 1968) be regarded as an estimation problem in a particular mixture distribution. The basic idea in latent class analysis is that the observed associations between a set of d dichotomous variates are generated by the presence of several different 'latent' classes within which the variables are independent. This may be formulated in mixture terms by supposing that a random vector $\boldsymbol{x}$ of dichotomous variates, arising from such a structure, has a probability density function given by

$$P(\boldsymbol{x}\,;\boldsymbol{p},\boldsymbol{\Theta}) = \sum_{j=1}^{c} p_j P_j(\boldsymbol{x}\,;\boldsymbol{\theta}_j) \quad (4.34)$$

$$P_j(\boldsymbol{x};\boldsymbol{\theta}_j) = \prod_{l=1}^{d} \theta_{jl}^{x_l}(1-\theta_{jl})^{1-xl} \tag{4.35}$$

where $\boldsymbol{p} = (p_1, \ldots, p_{c-1})$, $\boldsymbol{\Theta} = (\boldsymbol{\theta}_1, \ldots, \boldsymbol{\theta}_c)$, $x_l(l = 1, \ldots, d)$ are the elements of $\boldsymbol{x}$ and take values zero or one, and $\theta_{jl}(l = 1, \ldots, d)$ are elements of $\boldsymbol{\theta}_j$ and give the probability of the lth variable in class j being present (i.e. having a value of unity); c is the number of latent classes. By applying the general maximum likelihood equations for a finite mixture given in Chapter 1 to (4.34) and (4.35), we obtain the now familiar estimation equations

$$\hat{p}_r = \frac{1}{n}\sum_{i=1}^{n} \hat{P}(r|\boldsymbol{x}_i) \tag{4.36}$$

$$\hat{\boldsymbol{\theta}}_r = \frac{1}{n\hat{p}_r}\sum_{i=1}^{n} \boldsymbol{x}_i \hat{P}(r|\boldsymbol{x}_i) \tag{4.37}$$

where, as previously,

$$\hat{P}(r|\boldsymbol{x}_i) = \frac{\hat{p}_r P_r(\boldsymbol{x};\hat{\boldsymbol{\theta}}_r)}{P(\boldsymbol{x},\hat{\boldsymbol{p}},\hat{\boldsymbol{\Theta}})} \tag{4.38}$$

Example

To illustrate how this approach can work in practice, it was applied

Table 4.6. *Maximum likelihood estimates for the parameters in a latent class problem.*

Starting values	*Number of iterations*	*Final values*
$p_1 = 0.33$		0.25
$p_2 = 0.33$		0.23
$\boldsymbol{\theta}_1 = [0.50, 0.50, 0.20, 0.30, 0.10]$	81	[0.52, 0.39, 0.16, 0.36, 0.00]
$\boldsymbol{\theta}_2 = [0.30, 0.20, 0.70, 0.60, 0.40]$		[0.14, 0.04, 0.78, 0.75, 0.36]
$\boldsymbol{\theta}_3 = [0.90, 0.70, 0.50, 0.10, 0.70]$		[0.84, 0.70, 0.48, 0.11, 0.61]
$p_1 = 0.25$		0.31
$p_2 = 0.25$		0.27
$\boldsymbol{\theta}_1 = [0.30, 0.40, 0.40, 0.50, 0.30]$	103	[0.16, 0.55, 0.37, 0.37, 0.35]
$\boldsymbol{\theta}_2 = [0.40, 0.10, 0.50, 0.40, 0.60]$		[0.52, 0.00, 0.52, 0.61, 0.27]
$\boldsymbol{\theta}_3 = [0.60, 0.50, 0.40, 0.30, 0.50]$		[0.99, 0.72, 0.46, 0.03, 0.56]

to 300 observations generated from the distribution given by (4.34) with $c = 3$, $d = 5$, and parameter values as follows:

$$
\begin{aligned}
&p_1 = 0.33 \qquad p_2 = 0.33 \\
&\boldsymbol{\theta}_1 = [0.5, \quad 0.5, \quad 0.2, \quad 0.3, \quad 0.1] \\
&\boldsymbol{\theta}_2 = [0.3, \quad 0.2, \quad 0.7, \quad 0.6, \quad 0.4] \\
&\boldsymbol{\theta}_3 = [0.9, \quad 0.7, \quad 0.5, \quad 0.1, \quad 0.7].
\end{aligned}
$$

The results are shown in Table 4.6, and indicate that the choice of starting values is likely to have a considerable effect on the final estimates obtained. Nevertheless the 'mixture' approach to latent class analysis may, in many cases, present a viable alternative to the procedures generally adopted for this technique, i.e. those outlined by Gibson (1959) and others. □

4.6 Summary

This chapter has concentrated largely on finite mixtures involving binomial and Poisson distributions. However, the estimation procedures described could easily be adapted for mixtures of other discrete distributions, if these were of particular interest. These estimation techniques should, in many cases, be able to provide reasonable parameter estimates, but the previously issued caveats still apply, namely that large samples may be required in the case when component distributions are not well separated, and that, when using maximum likelihood, good initial estimates are usually necessary. Even when these are provided, convergence of the iterative algorithm described can be slow, as evidenced by the Poisson mixture example given in Section 4.3.2. Because of this, other iteration schemes for solving the maximum likelihood equations could be considered, for example the Newton–Raphson algorithm described in Chapter 2; however, the simplicity of the algorithm used throughout this chapter may, despite its possible slow convergence rate, make it more attractive to most users.

CHAPTER 5

Miscellaneous topics

5.1 Introduction

In this chapter we shall discuss a number of outstanding topics concerned with mixture distributions. The first will be the important and often difficult problem of determining the number of components in the mixture. The second will be the techniques of probability density function estimation in the context of finite mixtures. Finally, a number of miscellaneous points such as the generation of random variables from mixture distributions will be covered.

5.2 Determining the number of components in a mixture

In the previous chapters it has been implicitly assumed that c, the number of components in the mixture, is known before the estimation of parameters is attempted. In many cases this will be so, since the decision to fit a mixture distribution will be based upon theoretical knowledge of the application at hand, for example the known existence of two or more species, the presence of two sexes, etc. There are, however, situations where the decision to apply a mixture distribution must be based upon, or at least supported by, the sample data, and so questions arise as to the appropriate value of c, and in particular whether $c = 1$, in which case such a distribution would be unnecessary. Several methods have been proposed which may be helpful in such situations, particularly where mixtures of normal distributions are being considered. These techniques fall fairly naturally into two classes; the first contains informal graphical techniques such as examination of sample histograms, etc.; the second includes the more formal hypothesis testing variety of technique.

5.2.1 *Informal diagnostic tools for the detection of mixtures*

Perhaps the most natural candidate to consider when assessing whether a mixture distribution might be appropriate for univariate data is the sample histogram, and the most obvious feature of this that might be indicative of a mixture would be some form of multimodality. However, the mixing of two or more unimodal frequency curves will produce a multimodal distribution only under certain circumstances (see Chapter 2 for conditions for bimodality of a mixture of two normal distributions); since the fundamental property of interest is the mixing not the presence of more than a single mode, it is natural to look for other descriptive features likely to indicate the mixing of components. One such property is the existence of more than two points of inflexion, i.e. points at which the curve changes from being convex to concave. An equivalent idea is that of the curve being bitangential, i.e. of there being a line, other than the x-axis, that is tangent to the frequency curve at more than one point.

However, all such properties may be difficult to detect visually even from the population frequency curve, and it will obviously be still more difficult to draw conclusions from sample data. An illustration that this is so is given in Murphy (1964), who shows histograms of a number of samples of size 50 generated from a *single* normal distribution. Many of the histograms might, on visual inspection, be taken to show signs of bimodality or even trimodality. Of course, a formal test of multimodality such as that proposed by Haldane (1952) might usefully be applied in such situations, but Cox (1966) suggests that such tests are quite insensitive except in regions where the frequencies are extremely large.

Examination of the sample histogram is therefore unlikely to be of much help in detecting the presence of a mixture – indeed, it might prove positively misleading. Consequently we must consider other possibilities, and for examining data for the presence of mixtures of normal distributions a variety of techniques based on the familiar method of *probability plotting* have been suggested. Such plots have a long and reasonably honourable history in statistics, first being suggested by Hazen (1914) in connection with a study of floods. They have been used in a variety of ways, but their main application has been to obtain a quick informal check on distributional assumptions in the light of the sample and to obtain rough estimates of scale and location parameters. (Probability plotting is described in detail in Everitt, 1978.) For detecting mixtures of normal distributions the simplest type of probability plot which might be useful is that of the sample quantiles against those of a standard normal curve. If a single normal distribution is appropriate for the data such a plot should be approximately linear. In the case of a mixture the curve will be to some degree S-shaped, the extent of the departure from linearity depending on the separation of the components in the mixed distribution. To illustrate how this procedure operates in practice, it was applied to a variety of two-component normal mixture distributions with $\sigma_1 = \sigma_2 = 1.0$, $p = 0.5$, $\mu_1 = 0.0$, and varying values of μ_2. Sample sizes of 200 were used and the resulting plots are shown in Fig. 5.1 to 5.5. (The first of these shows the plot of 200 observations from a *single* normal distribution and is included for comparative purposes.) The departure from linearity becomes pronounced only for values of $\mu_2 \geqslant 3.0$, and so it would seem that such probability plots are, at best, relatively insensitive indicators of the presence of a mixture.

In a recent paper, Fowlkes (1979) describes a further probability plotting technique which he claims is more sensitive to the presence of a normal mixture than the quantile–quantile plot just outlined. His suggestion is to plot standardized sample quantiles, $x_{(i)}$, against the quantities

$$y_i = \Phi[x_{(i)}] - b_i \tag{5.1}$$

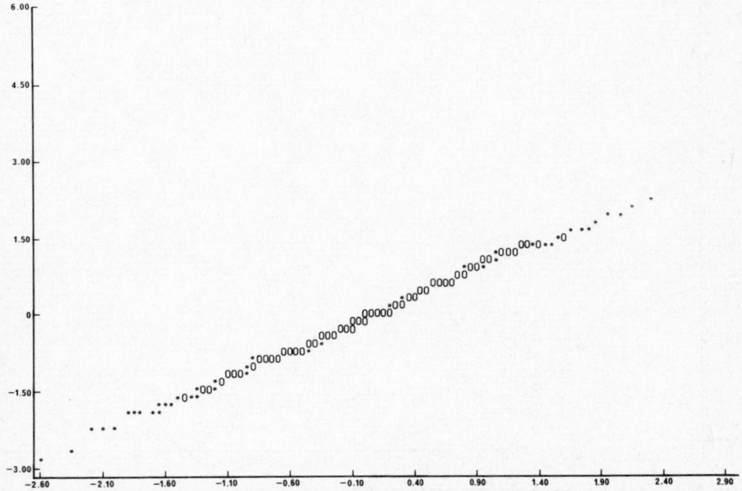

Fig. 5.1. *Normal probability plot of 200 observations from a single normal distribution. (In this and in Fig. 5.2 to 5.5, the theoretical quantiles are plotted along the x-axis, and the ordered observations along the y-axis.)*

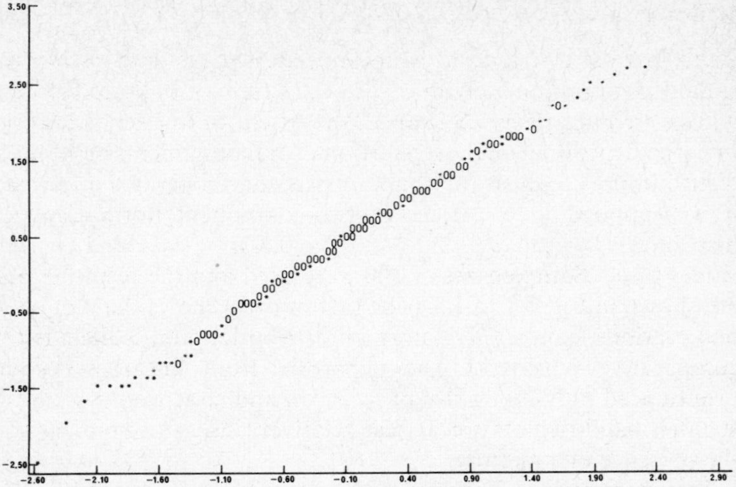

Fig. 5.2. *Normal probability plot of 200 observations from a mixture of two normal distributions with* $p = 0.5$, $\mu_1 = 0.0$, $\sigma_1 = 1.0$, $\mu_2 = 1.0$ *and* $\sigma_2 = 1.0$.

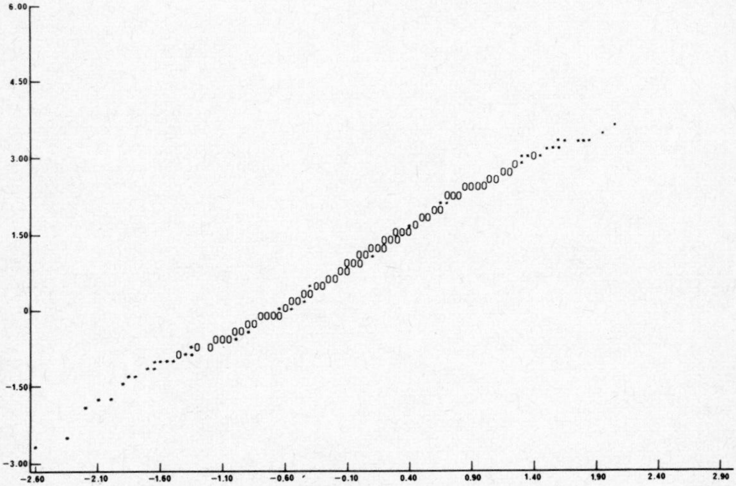

Fig. 5.3. *Normal probability plot of 200 observations from a mixture of two normal distributions with* $p = 0.5$, $\mu_1 = 0.0$, $\sigma_1 = 1.0$, $\mu_2 = 2.0$, $\sigma_2 = 1.0$.

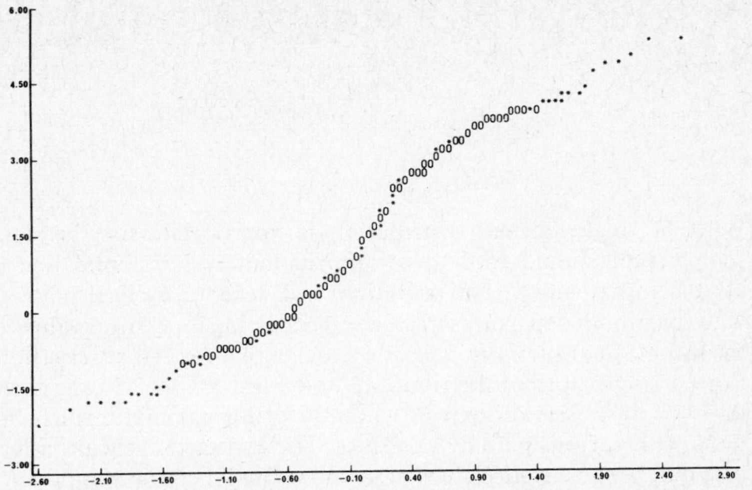

Fig. 5.4. *Normal probability plot of 200 observations from a mixture of two normal distributions with* $p = 0.5$, $\mu_1 = 0.0$, $\sigma_1 = 1.0$, $\mu_2 = 3.0$, $\sigma_2 = 1.0$.

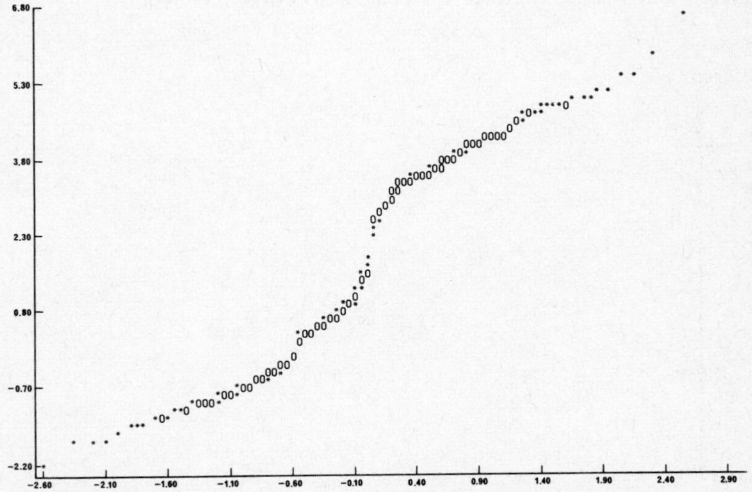

Fig. 5.5. *Normal probability plot of 200 observations from a mixture of two normal distributions with* $p = 0.5$, $\mu_1 = 0.0$, $\sigma_1 = 1.0$, $\mu_2 = 4.0$, $\sigma_2 = 1.0$.

where

$$\Phi(Z) = \int_{-\infty}^{Z} \frac{1}{\sqrt{(2\pi)}} \exp(-\tfrac{1}{2}u^2)\mathrm{d}u$$

$$b_i = (i - \tfrac{1}{2})/n,$$

When a single normal distribution is appropriate for the data, such a plot should lead to an approximately horizontal line at $y = 0$. Normal mixture distributions will lead to cyclical patterns with maximum departures from $y = 0$ occurring for central values of the sample quantiles. Fig. 5.6 shows such a plot for 200 observations from a single normal distribution, and Figs. 5.7 to 5.10 show the plots for 200 observations from a number of normal mixture distributions with various parameter values. The expected cyclical pattern becomes very clear for values of $\mu_2 \geqslant 3.0$, but when the components of the mixture are less well separated it becomes less obvious that departures from the line $y = 0$ are greater than in Fig. 5.6. Fowlkes makes the suggestion that in order to reduce local variation and to recover the cyclical pattern the plots might be smoothed by the methods described in Cleveland and Kleiner (1975). From the diagrams presented in his paper this does appear to improve the sensitivity of the method. Fowlkes also considers whether his plots

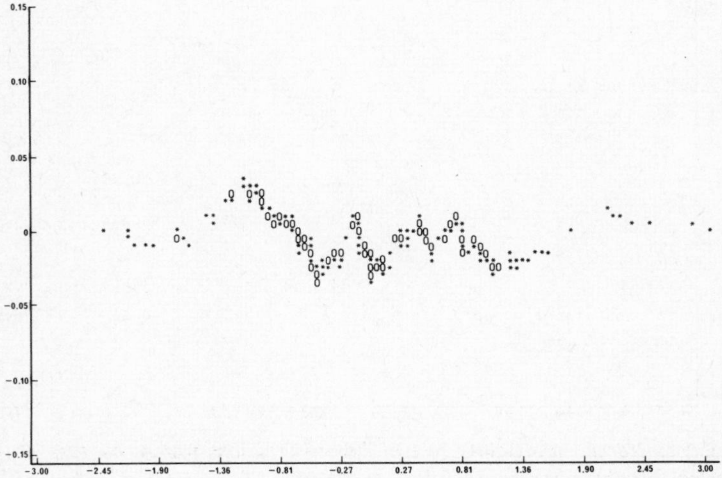

Fig. 5.6. *Fowlkes probability plot of 200 observations from a single normal distribution. (Axes as indicated in the text.)*

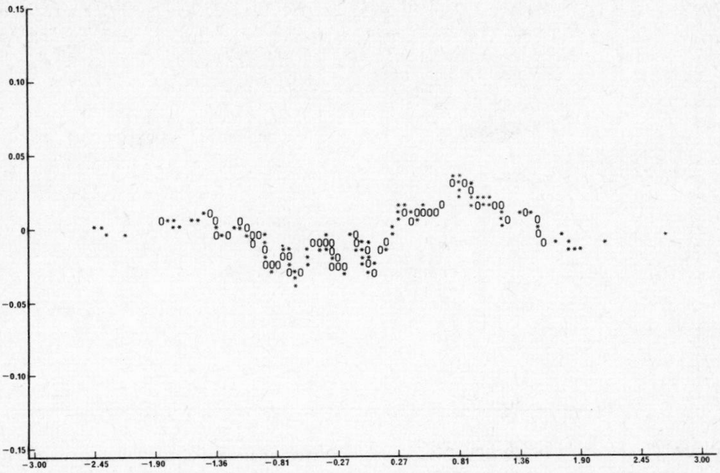

Fig. 5.7. *Fowlkes probability plot of 200 observations from a mixture of two normal distributions with* $p = 0.5$, $\mu_1 = 0.0$, $\sigma_1 = 1.0$, $\mu_2 = 1.0$, $\sigma_2 = 1.0$.

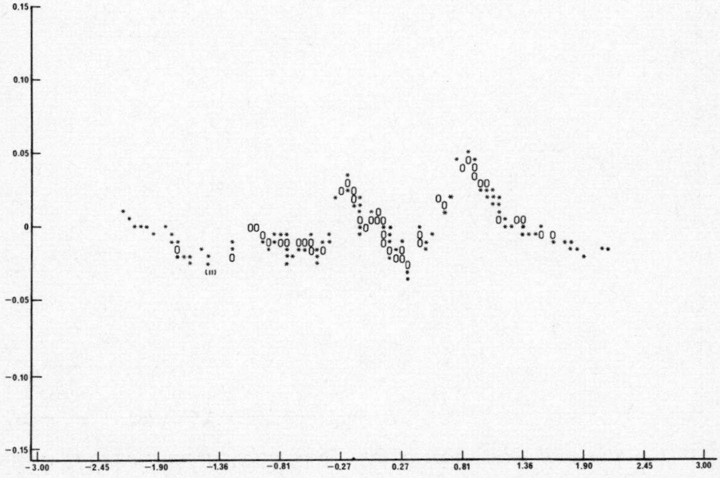

Fig. 5.8. *Fowlkes probability plot of 200 observations from a mixture of two normal distributions with* $p = 2.5$, $\mu_1 = 0.0$, $\sigma_1 = 1.0$, $\mu_2 = 2.0$, $\sigma_2 = 1.0$.

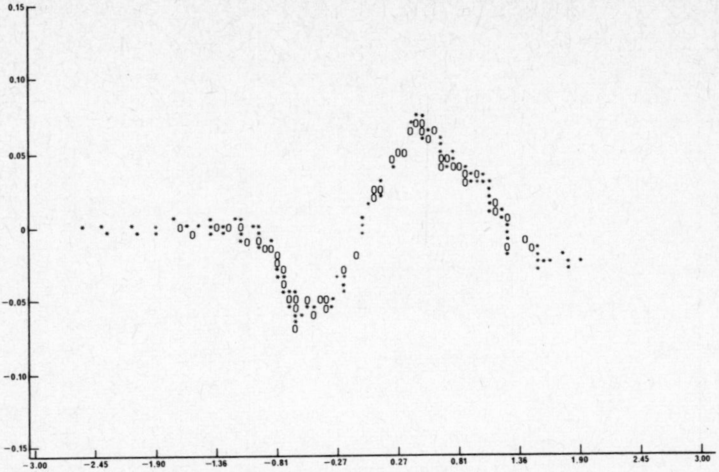

Fig. 5.9. *Fowlkes probability plot of 200 observations from a mixture of two normal distributions with* $p = 0.5$, $\mu_1 = 0.0$, $\sigma_1 = 1.0$, $\mu_2 = 3.0$, $\sigma_2 = 1.0$.

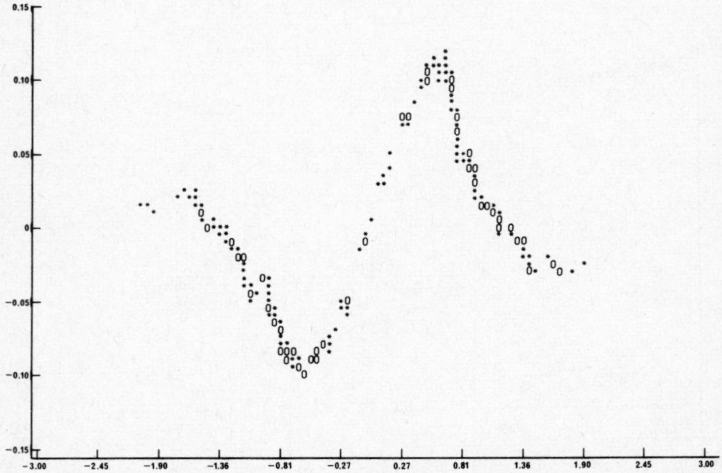

Fig. 10. *Fowlkes probability plot of 200 observations from a mixture of two normal distributions with* $p = 0.5$, $\mu_1 = 0.0$, $\sigma_1 = 1.0$, $\mu_2 = 4.0$, $\sigma_2 = 1.0$.

when produced by lack of normality can be distinguished from those produced by a mixture, and in an unpublished Bell Laboratories memorandum Fowlkes demonstrates that the plots for mixtures can be distinguished from those for simple non-normal distributions such as the gamma, t, and log–normal in most instances, even with relatively small samples.

For the detection of multivariate normal mixtures a possible aid is a chi-square probability plot of the generalized distance, D_i, of each observation from the sample mean vector, where

$$D_i = (\boldsymbol{x}_i - \bar{\boldsymbol{x}})\boldsymbol{S}^{-1}(\boldsymbol{x}_i - \bar{\boldsymbol{x}}) \tag{5.2}$$

and $\boldsymbol{S}$ is the sample variance–covariance matrix. If the data are from a single multivariate normal distribution these distances have, approximately, a chi-square distribution with d degrees of freedom. Consequently a chi-square probability plot of the ordered distances should result in a straight line through the origin. Mixtures of multivariate normals will tend to give S-shaped curves. Fig. 5.11 shows such a plot for a sample of 200 observations from a single multivariate normal population, and Fig. 5.12 shows the plot obtained from the iris setosa and iris versicolour observations of Fisher's iris data (see Chapter 2 for details). The latter plot shows clear evidence of departure from linearity and could be used to indicate that a normal mixture might be a possibility for these data.

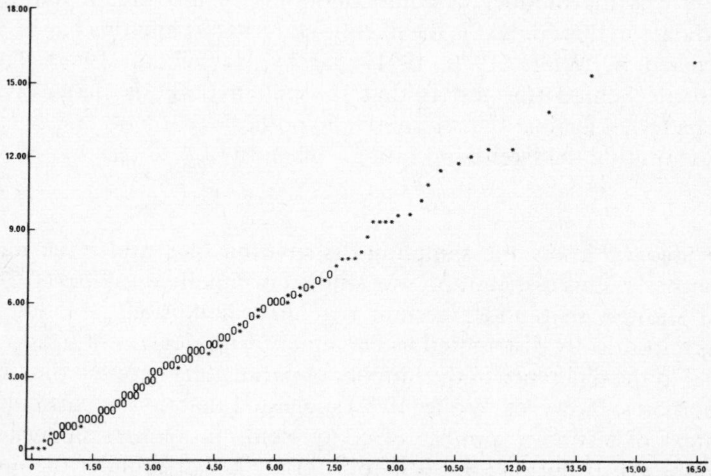

Fig. 5.11. *Chi-square probability plot of generalized distances for 200 observations from a single four-dimensional multivariate normal distribution.*

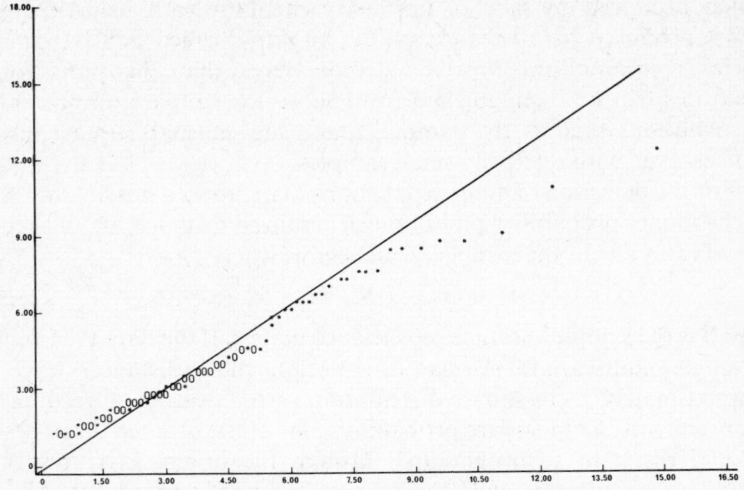

Fig. 5.12. *Chi-square probability plot of generalized distances for iris setosa and iris versicolour.*

5.2.2 *Testing hypotheses on the number of components in a mixture*

In this section we will consider the more formal methods which have been proposed for testing the hypothesis $c = c_0$ against $c = c_1$, where c is the number of components in the mixture. A natural candidate in this context is the likelihood ratio test and this has been discussed by Wolfe (1970, 1971) and by Hasselblad (1969). The rationale behind this test is that the null hypothesis that $c = c_0$ can be tested against the alternative hypothesis that $c = c_1 (c_1 > c_0)$ by computing the likelihood ratio, λ, given by

$$\lambda = \mathscr{L}_{c0} / \mathscr{L}_{c_1} \tag{5.3}$$

provided we know the sampling distribution of λ under the null hypothesis. This distribution was studied originally by Wilks (1938) who showed that under certain regularity conditions $-2 \log_e \lambda$ is asymptotically distributed as chi-square with degrees of freedom equal to the difference in the number of parameters between the two hypotheses. However, Wolfe (1971) suggested that, in the particular context of testing a number of components in a mixture, such a sampling distribution for λ is inappropriate because, under the null hypothesis, the mixing proportions $p_{c0+1}, p_{c0+2}, \ldots, p_{c_1}$, lie on the boundary of the parameter space (under H_0 they are all zero), so that regularity conditions for $-2\log\lambda$ to be asymptotically chi-squared are not fulfilled. (Such a point is also made by Binder, 1978.) For testing the number of components in a mixture of multivariate normals, Wolfe therefore suggests a modified test in which

$$-\frac{2}{n}\left(n-1-d-\frac{c_1}{2}\right)\log_e\lambda \tag{5.4}$$

is tested as a chi-square with degrees of freedom $2d(c_1 - c_0)$. Some simulation results of Everitt (1981) show that this suggestion is reasonable only in cases where $n > 10\,d$; in the same paper some power curves of the test are derived again by simulation, and these indicate that the test has fairly low power until the generalized distance between the two components is greater than two. Hasselblad (1969) uses the likelihood ratio test with mixtures of exponential, Poisson, and binomial distributions and finds its performance fairly satisfactory.

A number of authors have attempted to produce more specific tests for the presence of a mixture. For example, Johnson (1973) considers the problem of testing whether an observed sample is consistent with it being from a mixture (in unknown proportions) of two specified symmetrical populations, and suggests a test based upon the difference between the following two estimates of the mixing proportion, p:

$$\hat{p}_A = (\bar{x} - \mu_2)/(\mu_1 - \mu_2) \tag{5.5}$$

$$\hat{p}_B = (\bar{y} - P_2)/(P_1 - P_2) \tag{5.6}$$

where $\bar{x}$ is the sample mean, $\bar{y}$ is the proportion of sample values below some fixed value θ, and $P_j = \mathrm{pr}(x < \theta\,|\,\mu_j)$ $(j = 1, 2)$. If the two estimators $\hat{p}_A$ and $\hat{p}_B$ differ greatly then this may be regarded as evidence that the data are not distributed as a mixture of the two specified components. Johnson derives the variance of the difference $\hat{p}_A - \hat{p}_B$, for the particular case of normal components, and investigates the power of his suggested test against the alternative that the data arise from a single homogeneous normal distribution with a particular mean and variance.

Other authors who have considered this problem are Baker (1958), who suggests a test for detecting a two-component normal mixture which is based upon the statistic

$$D_j = \frac{1}{n}\sum_{i=1}^{n}\left|\frac{i}{n} - \Phi_i\right|^j \tag{5.7}$$

where

$$\Phi_i = \int_{-\infty}^{x_{(i)}} \frac{1}{\sqrt{(2\pi)}} \exp(-u^2/2)\mathrm{d}u$$

and $x_{(i)}$ is the standardized ith sample quantile; De Oliveira (1963), who describes a test for the presence of a Poisson mixture which is based upon the difference between the sample variance and mean; Binder (1978) who describes a number of tests appropriate for normal mixtures in the special case of testing $c = 1$ against $c = 2$. (A number of other relevant tests are described in Engleman and Hartigan, 1969, and Lee, 1979.) Although some of these tests might be useful in particular instances they are likely to have rather low power, since there are so many plausible kinds of departure from the hypothesis of a mixture with a specified number of components. Consequently it is unlikely that they will be helpful in general, and so for routine analysis of sample data for the presence of a mixture the likelihood ratio test described above is perhaps preferable (but see comments in the summary at the end of this chapter).

Aside from the tests described above for the number of components in a finite mixture, there appears to have been little work on more fundamental hypothesis-testing questions, such as specifying that the parameters of a particular finite mixture take on certain values. A major problem here in constructing exact parametric tests is that the derivation of the small-sample distribution of an estimator will, in general, be extremely difficult. Consequently it is desirable to use estimators having acceptable asymptotic properties, since we cannot make a choice on the basis of their small-sample behaviour; where estimators are, for example, asymptotically normally distributed, we can use the elements of the covariance matrix of the asymptotic distribution to construct confidence interval estimates and hence acceptance regions for testing hypotheses of a particular structure. (It should be emphasized that the question of identifiability is a crucial one in any hypothesis-testing problem; before the construction of a test procedure is attempted it must be verified that the alternative hypothesis is identifiable.)

5.3 Probability density function estimation

The emphasis in this book has been on estimating the parameters of a mixture distribution when there have been theoretical reasons for supposing that the population distribution was in fact a mixture.

However, one can also view a mixture as being one way of introducing extra parameters into an approximating function without necessarily supposing a mixture to be an appropriate model for the phenomenon. The introduction of more parameters would, of course, be expected to yield a better approximation. Approximating the population by a single distribution would be the simplest special case. Next we would have a mixture of two components, then three, and so on. This section outlines the approximating function which results when this process is continued until there are n components, one for each sample point in the mixture.

For simplicity we shall start with univariate distributions and begin with the general form for a mixture:

$$\hat{f}(x) = \sum_{i=1}^{c} p_i g_i(x; \boldsymbol{\theta}_i)$$

(where the notation $\hat{f}$ signifies that $\hat{f}$ is an estimate of f).

We set $c = n$ and $p_i = \frac{1}{n} (i = 1, \ldots, n)$, so that each component contributes equal weight. For further simplicity we suppose that each component has the same form, so that

$$g_i(x; \boldsymbol{\theta}_i) = g(x; \boldsymbol{\theta}_i) \qquad i = 1, \ldots, n$$

and

$$\hat{f}(x) = \frac{1}{n} \sum_{i=1}^{n} g(x; \boldsymbol{\theta}_i)$$

and we shall let $\boldsymbol{\theta}_i = (\mu_i, h_i)$, where μ_i is a location parameter and h_i is a spread parameter, and restrict g to a special form so that

$$\hat{f}(x) = \frac{1}{n} \sum_{i=1}^{n} \frac{1}{h_i} g\left(\frac{x - \mu_i}{h_i}\right).$$

Again for simplicity we shall let $h_i = h$ for all i. The μ_i, the location parameter for the ith component, will be estimated by $x_i (i = 1, \ldots, n)$, so that finally we have

$$\hat{f}(x) = \frac{1}{nh} \sum_{i=1}^{n} g\left(\frac{x - x_i}{h}\right). \tag{5.8}$$

This is a fairly general estimator for f. Now let us seek an intuitive justification for it. To do so, let us be quite specific and replace g by a normal component so that

$$\hat{f}(x) = \frac{1}{nh} \sum_{i=1}^{n} \frac{1}{\sqrt{(2\pi)}} \exp[-(x - x_i)^2/2h^2].$$

Now the estimated density at any point x is given by a sum of contributions, one for each of the sample points. The size of the contribution due to point x_i depends on the distance between x_i and x. The smaller this distance, the greater the contribution. Thus, in the

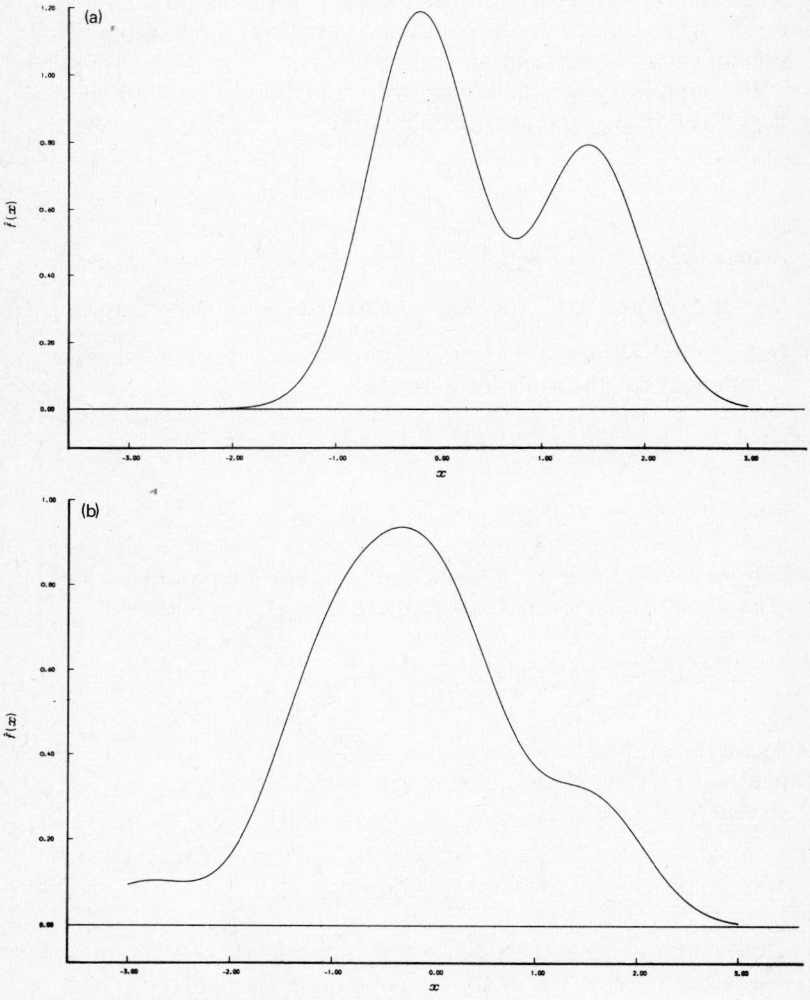

Fig. 5.13. (*Legend overleaf*)

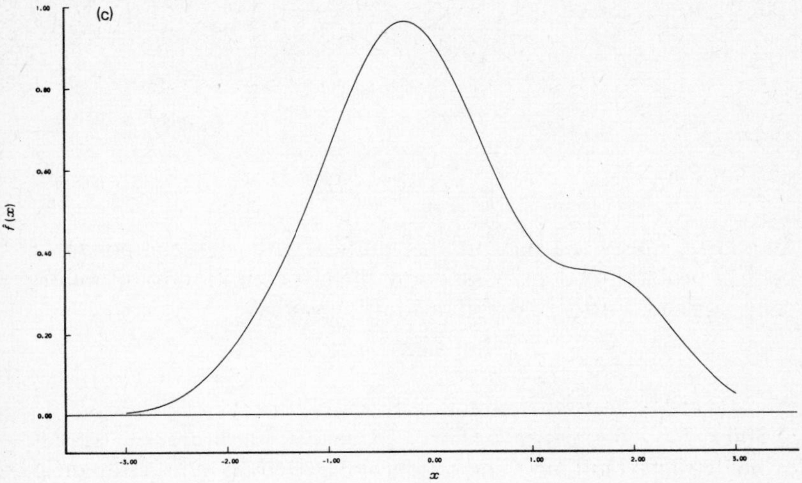

Fig. 5.13. *Kernel estimate for samples from a standard normal population with (a) $n = 5$, $h = 0.5$, (b) $n = 20$, $h = 0.5$ (c) $n = 50$, $h = 0.5$.*

vicinity of a compact cluster of sample points there will be several large contributions to the estimate which will consequently be larger. Fig. 5.13 shows $\hat{f}$ for several different sample sizes from a standard normal distribution.

This non-parametric probability density function estimator is known as the *kernel* estimator (with each component being a *kernel*) or the *Parzen estimator* after Parzen (1962) who first investigated the general form of the estimator. Earlier, Rosenblatt (1956) had studied the special case with

$$g\left(\frac{x-\mu}{h}\right) = \begin{cases} 0 & \text{if } |x-\mu| > h \\ \frac{1}{2} & \text{if } |x-\mu| \leqslant h \end{cases}.$$

Parzen gave conditions on h and g for the estimator to possess certain desirable statistical properties. For example, $\hat{f}(x)$ in (5.8) is asymptotically unbiased at all points x at which $f(x)$, the true probability density function, is continuous if

$$\lim_{n \to \infty} h(n) = 0 \tag{5.9}$$

$$\sup_{-\infty < z < \infty} |g(z)| < \infty$$

$$\int_{-\infty}^{\infty} |g(z)|\,\mathrm{d}z < \infty$$

$$\lim_{z\to\infty} |zg(z)| = 0$$

and

$$\int_{-\infty}^{\infty} g(z)\,\mathrm{d}z = 1.$$

Of course, since $\hat{f}$ is a mixture distribution with each component g being a probability density function, the last condition is automatically satisfied. Furthermore, if in addition we have

$$\lim_{n\to\infty} nh(n) = \infty$$

then $\hat{f}(x)$ is a consistent estimator of $f(x)$.

Since Parzen's paper a fairly extensive literature on kernel estimates has built up. The reader interested in non-parametric density function estimation in general is referred to Wertz (1978) and Wegman (1972).

So far, although we have described the role of h as that of a spread parameter, and although we have shown how it should change as n increases, we have said nothing about its absolute magnitude. Fig. 5.14, showing the results of a kernel estimate on a sample of

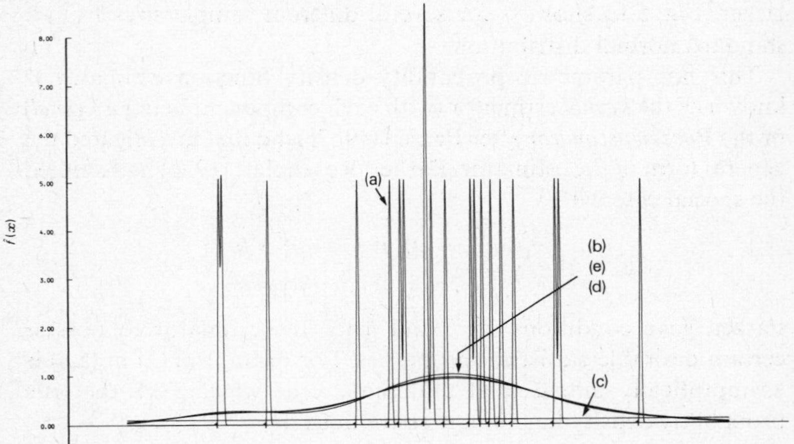

Fig. 5.14. *Kernel estimate for a sample of 20 points from a standard normal population with (a) $h = 0.01$, (b) $h = 0.5$, (c) $h = 10.0$, (d) $h = 0.62$, (e) $h = 0.58$.*

20 points from a univariate standardized normal distribution, illustrates that this is an important problem. Fig. 5.14(a) has $h = 0.01$, (b) has $h = 0.5$, (c) has $h = 20.0$. Clearly 0.01 is too small and 20.0 is too large. But how should we choose a good value, especially in the multivariate case to be discussed below, where the resulting estimate cannot be plotted? Several approaches have been suggested and we shall outline two.

The first begins by finding the average distance between sample points and their *m*th nearest sample point. If for the case of normal kernels discussed above this average distance is used as *h*, then our kernels are such that on average *m* points lie within one standard deviation of *x*. A value of $m = 10$ has been suggested as reasonable. Fig. 5.14(d) shows the result when this *h* value is applied to the above data.

The second method uses a jack-knife approach based on the likelihood function

$$L(h) = \prod_{i=1}^{n} f_1(x_i)$$

where $f_1(x_i)$ is the estimate at x_i based on the sample $\{x_1, x_2, \ldots, x_{i-1}, x_{i+1}, \ldots, x_n\}$. Fig. 5.14(e) shows this result.

A poor choice of kernel shape could also have a deleterious effect on the resulting estimate and investigations have been made to find the best shape. Indeed Parzen (1962) lists seven possible shapes. In applications the most popular seems to be the normal distribution. Its major possible disadvantage is that it has an unbounded region of support so that all sample points contribute something, no matter how little, to the estimate at every *x*. If speed of estimation is important then it might be appropriate to use a modified normal kernel, say by truncating it at three standard deviations.

A general kernel estimate in *d* dimensions is

$$f(\boldsymbol{x}) = \frac{1}{n} \sum_{i=1}^{n} g(\boldsymbol{x}\,; \boldsymbol{\theta}_i)$$

where g is a function of multivariate $\boldsymbol{x}$. We can immediately extend the univariate case to the multivariate by substituting the appropriate g. Thus, for normal kernels,

$$\hat{f}(\boldsymbol{x}) = \frac{1}{n} \sum_{i=1}^{n} \frac{1}{(2\pi)^{d/2} |\Sigma|^{1/2}} \exp\left[-\tfrac{1}{2}(\boldsymbol{x} - \boldsymbol{x}_i)' \Sigma^{-1} (\boldsymbol{x} - \boldsymbol{x}_i)\right]$$

where we have used a common Σ and where $\boldsymbol{\mu}_i$ is given by $\boldsymbol{x}_i (i = 1, \ldots, n)$. If this fairly general form is adopted then it has been suggested that Σ should be estimated by the variance–covariance matrix of the entire distribution. Frequently, however, the amount of calculation involved in the above $\hat{f}(\boldsymbol{x})$ is excessive and Σ is replaced by a diagonal matrix:

$$\hat{f}(\boldsymbol{x}) = \frac{1}{n} \sum_{i=1}^{n} \frac{1}{h_1 \ldots h_d (2\pi)^{d/2}} \prod_{j=1}^{d} \exp\left[-\frac{1}{2} \left(\frac{x_j - x_{ij}}{h_j} \right)^2 \right].$$

Sometimes even more simplification is made, setting $h_j = h (j = 1, \ldots, d)$, so that

$$\hat{f}(\boldsymbol{x}) = \frac{1}{nh^d (2\pi)^{d/2}} \sum_{i=1}^{n} \prod_{j=1}^{d} \exp\left[-\frac{1}{2} \left(\frac{x_j - x_{ij}}{h} \right)^2 \right].$$

For obvious reasons such a multivariate kernel is called a *product kernel*. Cacoullos (1966) was one of the first to investigate multivariate kernel estimates.

5.4 Miscellaneous problems

In this section we present miscellaneous problems concerning mixtures which do not merit a section of their own.

Bignami and de Matteis (1971) address the problem of sampling from a probability density function which is a linear combination of components. If the coefficients are positive we have a mixture distribution and the problem is straightforward. We simply select a component $g_i(x)$ with probability proportional to its weight p_i and then randomly select an x from $g_i(x)$. If the p_i can go negative the problem is more complicated. Consider first the two-component case

$$f(x) = p_1 g_1(x) - p_2 g_2(x)$$

with $p_1, p_2 > 0$.

Bignami and Matteis suggest the following solution. First select an x from the component $g_1(x)$. Then choose a random number r from the uniform distribution on $(0, 1)$ and accept it if $r \leqslant f(x)/p_1 g_1(x)$. If x is not accepted, select another x from $g_1(x)$ and choose another r.

The first step has a probability density function $g_1(x)$ and the second gives a probability $f(x)/p_1 g_1(x)$ of retaining the chosen x.

Since the two steps are independent the overall probability density function is proportional to the product

$$g_1(x)\frac{f(x)}{p_1 g_1(x)} \propto f(x).$$

In general, suppose

$$f(x) = \sum_{i=1}^{c} p_i g_i(x) - \sum_{j=1}^{d} q_j g_j(x)$$

with all p_i and q_j positive. Then the first step is to choose x from the p.d.f.

$$\sum_{i=1}^{c} p_i g_i(x) \Big/ \sum_{i=1}^{c} p_i$$

(where the denominator is simply a constant to yield unit integral). The second step is to generate r from the uniform distribution and accept x if

$$r \leqslant f(x) \Big/ \sum_{i=1}^{c} p_i g_i(x).$$

Again the overall probability density of x is proportional to

$$\frac{\sum_{i=1}^{c} p_i g_i(x)}{\sum_{i=1}^{c} p_i} \frac{f(x)}{\sum_{i=1}^{c} p_i g_i(x)} \propto f(x).$$

We dismissed above, as straightforward, the problem of sampling from a formal mixture distribution (with all weights positive), but its simplicity does not imply a lack of importance. Pseudo-random numbers generated by computer have application in a very diverse range of areas. Usually numbers are initially generated from a uniform distribution and some transformation is applied so that they follow the desired distribution. Often, however, it is important to have a very fast number generator and this method may not be fast enough. With this in mind, Marsaglia (1961) presented a general method for which the average time taken to generate a number is very small. Suppose we wish to generate numbers from the distribution $f(x)$. We first divide $f(x)$ in the form

$$f(x) = p_1 g_1(x) + p_2 g_2(x)$$

where $g_1(x)$ is of a form from which it is easy to generate random samples (such as the uniform distribution) and where $p_1 \gg p_2$. Then with high probability p_1 we sample from $g_1(x)$ and with low probability p_2 we sample from (the possibly complicated density) $g_2(x)$. The relative infrequency with which a sample from $g_2(x)$ is required results in a small average generation time.

The idea can be extended to

$$f(x) = \sum_{i=1}^{c-1} p_i g_i(x) + p_c g_c(x)$$

where $p_1 > p_2 > \ldots > p_{c-1} > p_c$ and where each of the $g_i(x)(i = 1, \ldots, c-1)$ have a simple form (usually uniform). For simplicity it is usual to keep

$$f(x) - \sum_{i=1}^{c-1} p_i g_i(x) \geqslant 0.$$

On a completely different note, Silverman (1978) presents a method for estimating the ratio $f(x)/g_1(x)$, where $f(x)$ is a mixture of $g_1(x)$ and some other unknown component $g_2(x)$, based on independent samples from $f(x)$ and $g_1(x)$. From this he derives a graphical method for estimating the mixing proportion and gaining an understanding of the form of the second component $g_2(x)$. The approach is based on the following idea.

Let

$$f(x) = p_1 g_1(x) + p_2 g_2(x).$$

Then a plot of $f(x)/g_1(x)$ against x will be approximately constant for small values of $g_2(x)/g_1(x)$ and this constant value will be an estimate of p_1. Furthermore, the part of the curve where $g_2(x)/g_1(x)$ is not small will indicate the shape of $g_2(x)/g_1(x)$.

As stated in the introduction, this book deals solely with finite mixtures since these have the greater practical importance. However, it would be an oversight not to include some mention of the theoretical results and practical applications of infinite mixtures. The references which follow are in no sense intended to be an exhaustive list of the studies of non-finite mixtures. They are merely intended as an introduction to the relevant literature.

Tallis (1969) has studied the identifiability of mixtures, beginning with countably infinite mixtures, with distributions taking the form

$$F(x) = \sum_{i=1}^{\infty} p_i G_i(x)$$

$\left(\text{with } \sum_{i=1}^{\infty} |p_i| < \infty \text{ and } \sum_{i=1}^{\infty} p_i = 1\right)$, and then moving on to discuss the general case

$$F(x) = \int_{-\infty}^{\infty} F(x, \theta) \mathrm{d}P(\theta).$$

Blum and Susarla (1977) also study the general case, concerning themselves with estimation of the mixing function $P(\theta)$. Deely and Kruse (1968) have also studied this.

Blischke (1963), though chiefly concerned with finite mixtures, gives parameter estimates for several common forms of infinite mixture.

As mentioned in Chapter 3, Padgett and Tsokos (1978) and Harris and Singpurwalla (1968) consider mixtures with continuous functions as mixing distributions.

Many of the more theoretical papers referred to in preceding chapters have results which apply to non-finite mixtures as well as to finite mixtures (see, for example, Teicher, 1961).

One area of practical interest where infinite mixtures have been applied is that of contagion (see for example, Feller, 1943, referred to by Blischke, 1963, and Chatfield and Theobald, 1973).

Finally, it would be a mistake not to point out the formal similarity between the general mixture distribution above and the Bayes estimate of the distribution of x.

5.5 Summary

Finite mixture distributions arise in many practical situations, a number of which have been illustrated in this text. Many methods have been suggested for estimating their parameters and these have been discussed in some detail in the previous chapters. The great majority of the estimation techniques assume that the number of components, c, in the mixture is known *a priori*, and this chapter has discussed various methods which have been suggested for testing hypotheses concerning c. However, no really adequate tests are available and this may be a consequence of the problem rather than any lack of ingenuity; consequently we should perhaps conclude by reminding readers that given enough components we can always find a mixture that 'fits' a set of data, and so, in general, the decision to fit such a distribution should be based upon some *sound theoretical reasons* rather than possible idle speculation.

References

Anderson, J.A. (1979), Multivariate logistic compounds. *Biometrika*, **66**, 17–26.

Antoniak, C.E. (1974), Mixtures of Dirichlet processes with applications to Bayesian nonparametric problems. *Ann. Stat.*, **2**, 1152–74.

Apolloni, B. and Murino, P. (1979), On the identification of a mixture of samples from two multivariate distributions. *Acta Astronautica*, **6**, 1031–42.

Ashford, J.R. and Hunt, R.G. (1974), The distribution of doctor-patient contacts within the NHS. *J. Roy. Statist. Soc., Series A*, **137**, 347–83.

Ashford, J.R. and Walker, P.J. (1972), Quantal response analysis for a mixture of populations. *Biometrics*, **28**, 981–88.

Ashton, W.D. (1971), Distribution for gaps in road traffic. *J. Inst. Maths. and its Applics.*,**7**, 37–46.

Baker, G.A. (1958), Empiric investigation of a test of homogeneity for populations composed of normal distributions. *J. Amer. Statist. Assoc.*, **53**, 551–57.

Barlow, R.E., Bartholomew, D.J., Bremner, J.M. and Brunk, H.D. (1972), *Statistical inference under order restrictions*, Wiley, London.

Bartholomew, D.J. (1959), Note on the measurement and prediction of labour turnover. *J. Royal Statist. Soc., Series A*, **122**, 232–39.

Bartholomew, D.J. (1969), Sufficient conditions for a mixture of exponentials to be a probability density function. *Annals of Math. Stat.*, **40**, 2183–88.

Bartlett, M.S. and MacDonald, P.D.M. (1968), Least Squares estimation of distribution mixtures. *Nature*, **217**, 195–96.

Behboodian, J. (1970), On a mixture of normal distributions. *Biometrika*, **57**, 215–17.

Behboodian, J. (1970), On the modes of a mixture of two normal distributions. *Technometrics*, **12**, 131–39.

Behboodian, J. (1972), On the distribution of a symmetric statistic from a mixed population. *Technometrics*, **14**, 919–23.

Behboodian, J. (1972), Information matrix for a mixture of two normal distributions. *J. Statist. Comp. Simul.*, **1**, 295–314.

Bhattacharya, C.G. (1967), A simple method of resolution of a distribution into Gaussian components. *Biometrics*, **23**, 115–35.

Bignami, A and De Matteis, A. (1971), A note on sampling from combinations of distributions, *J. Inst. of Maths and its Applics.*, **8**, 80–81.

Binder, D.A. (1978), Bayesian cluster analysis. *Biometrika*, **65**, 31–38.

Binder, D.A. (1978*a*), Comment on 'Estimating mixtures of normal distributions and switching regressions.' *J. Amer. Statist. Assoc.*, **73**, 746–47.

Blischke, W.R. (1962), Moment estimators for the parameters of a mixture of two binomial distributions. *Annals of Math. Stats.*, **33**, 444–54.

Blischke, W.R. (1963), Mixtures of discrete distributions. *Proc. Int. Symp. on classical and contagious discrete distributions.* Montreal, Pergamon Press, 351–72.

Blischke, W.R. (1964), Estimating the parameters of mixtures of binomial distributions. *J. Amer. Statist. Ass.*, **59**, 510–28.

Blischke, W.R. (1978), Mixtures of distributions. *International Encyclopedia of Statistics*, Edited by Kruskal, W.H. and Tanur, J.M., The Free Press, New York.

Blum, J.R. and Susarla, V. (1977), Estimation of a mixing distribution function. *Annals of Probability*, **5**, 200–209.

Blumenthal, S. and Govindarajulu, Z. (1977), Robustness of Stein's 2-stage procedure for mixtures of normal populations. *J. Amer. Statist. Assoc.*, **72**, 192–96.

Boes, D.C. (1966), On the estimation of mixing distributions. *Ann. Math. Statist.*, **37**, 177–88.

Bohrer, R. and Francis, G.K. (1972), Sharp one-sided confidence bounds over positive regions. *Ann. Math. Statist.*, **43**, 1541–48.

Bowman, K.O. and Shenton, L.R. (1973), Space of soultions for a normal mixture. *Biometrika*, **60**, 629–36.

Bowman, K.O. and Shenton, L.R. (1973), Remarks on the distribution of $\sqrt{b_1}$ in sampling from a normal mixture and normal type A distribution, *J. Amer. Statist. Assoc.*, **68**, 998–1003.

Bremner, J.M. (1978), Mixtures of Beta distributions–Algorithm AS 123. *Applied Statistics*, **27**, 104–09.

Brown, G.H. (1976), Combining estimates of category and subcategory proportions in a mixed population. *Biometrics*, **32**, 453–57.

Brown, G.H. and Fisher, N.I. (1972), Subsampling a mixture of sampled material. *Technometrics*, **14**, 663–68.

Brown, G.H. and Robson, D.S. (1975), The estimation of mixing proportions by double sample using bulk measurement. *Technometrics*, **17**, 119–26.

Bryant, P. and Williamson, J.A. (1978), Asymptotic behaviour of classification ML estimates. *Biometrika*, **65**, 273–81.

Burrau, C. (1934), The half-invariants of the sum of two typical laws of errors, with an application to the problem of dissecting a frequency curve into components. *Skand. Aktuarietidskr.*, **17**, 1–6.

Cacoullos, T. (1966), Estimation of a multivariate density. *Ann. of. the Inst. of Stat. Maths.*, **18**, 178–89.

Cassie, R.M. (1954), Some uses of probability paper in the analysis of size frequency distributions. *Austral. J. of Marine and Freshwater Res.*, **5**, 513–22.

Chandra, S. (1977), On the mixtures of probability distributions. *Scand. J. Statist.*, **4**, 105–12.

Chang, W.C. (1976), The effects of adding a variable in dissecting a mixture of two normal populations with a common covariance matrix. *Biometrika*, **63**, 676–78.

Chang, W.C. (1979), Confidence interval estimation and transformation of data in a mixture of two multivariate normal distributions with any given large dimension. *Technometrics*, **21**, 351–55.

Charlier, C.V.L. (1906), Researches into the theory of probability. *Lunds Universitets Årskrift, Ny följd*, **Afd. 2.1**, No. 5.

Charlier, C.V.L. and Wicksell, S.D. (1924), On the dissection of frequency functions. *Arkiv för Matematik, Astronomi och Fysik.*, **Bd. 18**, No. 6.

Chatfield, C. and Theobald, C.M. (1973), Mixtures and random sums. *Statistician*, **22**, 281–87.

Chien, Y.T. and Fu, K.S. (1967), On Bayesian learning and stochastic approximation. *IEEE Transactions in Systems Science and Cybernetics*, **SSC-3**, No. 1, 28–38.

Choi, K. and Bulgren, W.G. (1968), An estimation procedure for mixtures of distributions. *J. Royal Statist. Soc., Series B*, **30**, 444–60.

Choi, S.C. (1979), Two sample tests for compound distributions for homogeneity of mixing proportions, *Technometrics*, **21**, 361–65.

Clark, V.A., Chapman, J.M., Coulson, A.H. and Hasselblad, V. (1968), Dividing the blood pressures from the Los Angeles heart study into two normal distributions. *John Hopkins Medical Journal*, **122**, 77–83.

Cleveland, W.S. and Kleiner, B. (1975), A graphical technique for enchancing scatterplots with moving statistics. *Technometrics*, **17**, 447–54.

Cohen, A.C. (1960), An extension of a truncated Poisson distribution. *Biometrics*, **16**, 446–50.

Cohen, A.C. (1963), Estimation in mixtures of discrete distributions. *Proc. Int. Symp. on Classical and Contagious Discrete Distributions*. Montreal, Pergamon Press, 373–78.

Cohen, A.C. (1965), *Estimation in mixtures of Poisson and mixtures of exponential distributions*. NASA Technical Memorandum, NASA TM X-53245, George C. Marshall Space Flight Center, Huntsville, Alabama.

Cohen, A.C. (1966), A note on certain discrete mixed distributions. *Biometrics*, **22**, 566–72.

Cohen, A.C. (1967), Estimation in Mixtures of Two Normal Distributions. *Technometrics*, **9**, 15–28.

Cohen, A and Sackrowitz, H.B. (1974), On estimating the common mean of two normal distributions. *Ann. Statist.*, **2**, 1274–82.

Cooper, P.W. (1964), Non supervised adaptive signal detection and pattern recognition. *Information and Control*, **7**, 416–44.

Cox, D.R. (1962), *Renewal Theory*, Methuen and Co., London.

Cox, D.R. (1966), Notes on the analysis of mixed frequency distributions. *Br. J. Math. Statist. Psychol.*, **19**, 39–47.

Cox, D.R. and Hinkley, D.V. (1974), *Theoretical Statistics.* Chapman and Hall, London.

Craig, L.C. (1944), Identification of small amounts of organic compounds by distribution studies. II. Separation by counter-current distribution. *J. Biol. Chem.*, **155**, 519–34.

Davis, D.J. (1952), An analysis of some failure data. *J. Amer. Statist. Assoc.*, **47**, 113–50.

Dawid, A.P. and Skene, A.M. (1979), Maximum Likelihood Estimation of Observer Error-Rates using the EM algorithm. *Applied Statistics*, **28**, 20–28.

Day, N.E. (1969), Estimating the components of a mixture of normal distributions. *Biometrika*, **56**, 463–74.

Deely, J.J. and Kruse, R.L. (1968), Construction of sequences estimating the mixing distribution. *Ann. Math. Statist.*, **39**, 286–88.

Dempster, A.P., Laird, N.M. and Rubin, D.B. (1977), Maximum likelihood from incomplete data via the EM algorithm. *J. Royal Statist. Soc., Series B*, **39**, 1–38.

De Oliveira, J. (1963), Some elementary tests for mixtures of discrete distributions. *Proc. Int. Symp. on classical and contagious discrete distributions.* Montreal, Pergamon Press, 379–84.

Dick, N.P. and Bowden, D.C. (1973), Maximum likelihood estimation for mixtures of two normal distributions. *Biometrics*, **29**, 781–90.

Doetsch, G. (1928), Die elimination des dopplereffekts bei spektroskopischen feinstrukturen and exakte bestimming der komponenten. *Zeitschr. f. Phys.*, **49**, 705–30.

Doetsch, G. (1936), Zerlegung einer funktion in Gauβ'sche fehlerkurven und zeitliche zuruckverfolgung eines temperaturzustandes. *Math. Zeit.*, **41**, 283–318.

Duda, R. and Hart, P. (1973), *Pattern Classification and Scene Analysis.* J. Wiley and Son, New York.

Dynkin, E.B. (1951), *Selected translations in mathematical statistics and probability*. Printed for the Institute of Mathematical Statistics by the American Mathematical Society, 1961.

Eisenberger, I. (1964), Genesis of bimodal distributions. *Technometrics*, **6**, 357–63.

Engleman, L. and Hartigan, J.A. (1969), Percentage points of a test for clusters. *J. Amer. Statist. Assoc.*, **64**, 1647–48.

Epstein, B. and Sobel, M. (1953), Life testing. *J. Amer. Statist. Assoc.*, **48**, 486–502.

Everitt, B.S. (1978), *Graphical Techniques for Multivariate Data*, Heinemann, London.

Everitt, B.S. (1980), *Cluster Analysis*, 2nd Edition, Heinemann Educational Books, London.

Everitt, B.S. (1981), A Monte Carlo investigation of the likelihood ratio test for the number of components in a mixture of normal distributions. *Multiv. Behav. Res.*, in press.

Falls, L.W. (1970), Estimation of parameters in compound Weibull distributions. *Technometrics*, **12**, 399–407.

Feller, W. (1943), On a general class of 'contagious' distributions. *Annals of Math. Stats.*, **14**, 389–400.

Fowlkes, E.B. (1979), Some methods for studying the mixture of two normal (lognormal) distributions. *J. Amer. Statist. Assoc.*, **74**, 561–75.

Fryer, J.G. and Robertson, C.A. (1972), A comparison of some methods for estimating mixed normal distributions. *Biometrika*, **59**, 639–48.

Gargantini, I. and Henrici, P. (1967), See Cern Computer Centre Program Library, C202.

Gibson, W.A. (1959), Three multivariate models: Factor Analysis, Latent Structure Analysis and Latent Profile Analysis. *Psychometrika*, **24**, 229–52.

Green, B.F. (1951), A general solution for the latent class model of Latent Structure Analysis. *Psychometrika*, **16**, 151–66.

Gregor, J. (1969), An algorithm for the decomposition of a distribution into Gaussian components. *Biometrics*, **25**, 79–93.

Hald, A. (1948), *Statistiske metoder*, Copenhagen.

Hald, A. (1949), Maximum likelihood estimation of the parameters of a normal distribution which is truncated at a known point. *Skand. Aktuarietidskr.*, **32**, 119–34.

Hald, A. (1952), *Statistical theory with engineering applications*, John Wiley and Sons, New York.

Haldane, J.B.S. (1952), Simple tests for bimodality and bitangentiality. *Ann. Eugen.*, **16**, 359–64.

Harding, J.P. (1949), The use of probability paper for the graphical analysis of polymodal frequency distributions, *J. of the Marine Biol. Ass. of the UK*, **28**, 141–53.

Harris, C.M. and Singpurwalla, N.D. (1968), Life distributions derived from stochastic hazard functions. *Trans. IEEE on Reliability*, **17**, 70–79.

Harris, H. and Smith, C.A.B. (1949), The sib-sib age of onset correlation among individuals suffering from the same hereditary syndrome produced by more than one gene. *Ann. Eugen.*, **14**, 309–18.

Hasselblad, V. (1966), Estimation of parameters for a mixture of normal distributions. *Technometrics*, **8**, 431–44.

Hasselblad, V. (1969), Estimation of finite mixtures of distributions from the exponential family. *J. Amer. Statist. Assoc.*, **64**, 1459–71.

Hawkins, R.H. (1972), A note on multiple solutions to the mixed distribution problem. *Technometrics*, **14**, 973–76

Hazen, A. (1914), Storage to be provided in impounding reservoirs for municipal water supply. *Trans. Amer. Soc. Civil Engrs.*, **77**, 1539–659.

Hildebrand, F.B. (1956), *Introduction to numerical analysis*. McGraw-Hill, New York.

Hill, B.M. (1963), Information for estimating the proportions in mixtures of exponential and normal distributions. *J. Amer. Statist. Assoc.*, **58**, 918–32.

Holgersson, N. and Jorner, U. (1976), *The decomposition of a mixture into normal components: a review.* Research Report 76-13, University of Uppsala, Sweden.

Hosmer, D.W. (1973), On MLE of the parameters in a mixture of two normal distributions when the sample size is small. *Communications in Statistics*, **1**, 217–27.

Hosmer, D.W. (1973), A comparison of iterative maximum likelihood estimates of the parameters of a mixture of two normal distributions under three different types of sample. *Biometrics*, **29**, 761–70.

Hosmer, D.W. (1974), Maximum Likelihood estimates of the parameters of a mixture of two regression lines. *Communications in Statistics*, **3**, 995–1006.

Hosmer, D.W. (1978), Comment on Quandt and Ramsey's paper. *J. Amer. Statist. Assoc.*, **73**, 741–44.

Hyrenius, H. (1950), Distribution of 'Student'–Fisher t in samples from compound normal function. *Biometrika*, **37**, 429–42.

James, I.R. (1978), Estimation of the mixing proportion in a mixture of two normal distributions from simple, rapid measurements. *Biometrics*, **34**, 265–75.

Joffe, A.D. (1964), Mixed exponential estimation by the method of half moments. *Applied Statistics*, **13**, 91–98.

John, S. (1970), On identifying the population of origin of each observation in a mixture of observations from two gamma populations. *Technometrics*, **12**, 565–68.

Johnson, N.L. (1973), Some simple tests of mixtures with symmetrical components. *Communications in Statistics*, **1**, 17–25.

Johnson, N.L. and Kotz, S. (1970*a*), *Continuous Univariate Distributions-1*, Wiley, New York.

Johnson, N.L. and Kotz, S. (1970*b*), *Continuous Univariate Distributions-2*, Wiley, New York.

Kabir, A.B.M.L. (1968), Estimation of parameters of a finite mixture of distributions. *J. Royal Statist. Soc., Series B*, **30**, 472–82.

Kao, J.H.K. (1959), A graphical estimation of mixed Weibull parameters in life-testing electron tubes. *Technometrics*, **1**, 389–407.

Kazokos, D. (1977), Recursive estimation of prior probabilities using a mixture. *IEEE Trans. on Information Theory*, **IT-23**, 203–11.

Keilson, J. and Steutel, F.W. (1972), Families of infinitely divisible distributions closed under mixing and convolution. *Ann. Math. Statist.*, **43**, 242–50.

Kendall, M.G. and Stuart, A. (1963), *The Advanced Theory of Statistics* III, Griffin, London.

Kilpatrick, S.J. (1977), An empirical study of the distribution of episodes of illness recorded in the 1970-71 National Morbidity Survey. *Applied Statistics*, **26**, 26–33.

Krysicki, W. (1966), Zastosowanie metody momentów do estymacji parametrów mieszaniny dwóch rozkladów Laplace'a, *Zeszyty Naukowe Politechniki* Lódzkiej, **59**, 5–13.

Kumar, K.D., Nicklin, E.H. and Paulson, A.S. (1979), Comment on 'Estimating Mixtures of Normal Distributions and Switching Regression.'. *J. Amer. Statist. Assoc.*, **74**, 52–55.

Laird, N. (1978), Nonparametric maximum likelihood estimation of a mixing distribution. *J. Amer. Statist. Soc.*, **73**, 805–11.

Lazarsfeld, P.F. (1968), *Latent Structure Analysis*. Houghton Mifflin Co., Boston.

Lee, K.L. (1979), Multivariate Test for Clusters. *J. Amer. Statist. Assoc.*, **74**, 708–14.

MacDonald, P.D.M. (1971), Comment on 'An estimation procedure for mixtures of distributions' by Choi and Bulgren. *J. Royal Statist. Soc.*, *Series B*, **33**, 326–29.

MacQueen, J. (1967), Some methods for classification and analysis of multivariate observations. *Proc. 5th Berkeley Symp.*, **1**, 281–97.

Mardia, K.V. and Sutton, T.W. (1975), On the modes of a mixture of two von Mises distributions. *Biometrika*, **62**, 699–701.

Marriott, F.C. (1975), Separating mixtures of normal distributions. *Biometrics*, **31**, 767–69.

Marsaglia, G. (1961), Expressing a random variable in terms of uniform random variables. *Ann. Math. Stats.*, **32**, 894–98.

Martin, R.D. and Schwartz, S.C. (1972), On mixture, quasi-mixture and nearly normal random processes, *Ann. Math. Stats.*, **43**, 948–67.

Medgyessi, P. (1953), The decomposition of compound probability distributions. *Hungarian Acad. Sci. Comm. Inst. Appl. Math.*, **II**, 165–77.

Medgyessi, P. (1961), *Decomposition of superpositions of distribution functions*. Publishing house of the Hungarian Academy of Sciences, Budapest.

Meeden, G. (1972), Bayes estimation of the mixing distribution, the discrete case. *Ann. Math. Stats.*, **43**, 1993–99.

Mendenhall, W. and Hader, R.J. (1958), Estimation of parameters of mixed exponentially distributed failure time distributions from censored life test data. *Biomerika*, **45**, 504–20.

Morrison, D.F. (1971), Expectations and variances of maximum likelihood estimates of the multivariate normal distribution with missing data. *J. Amer. Statist. Soc.*, **66**, 602–604.

Muller, (1956), *MTAC*, 208–215. See Cern Computer Centre Program Library C 204.

Murphy, E.A. (1964), One Cause? Many Causes? The argument from the bimodal distribution. *J. Chron. Dis.*, **17**, 301–24.

Murphy, E.A. and Bolling, D.R. (1967), Testing of a single locus hypothesis where there is incomplete separation of the phenotypes. *Amer. J. Hum. Genetics*, **19**, 322–34.

Murray, G.D. and Titterington, D.M. (1978), Estimation problems with data from a mixture. *Applied Statistics*, **27**, 325–34.

Oka, M. (1954), Ecological studies on the kidai by the statistical method. II.

On the growth of kidai (Taitus tumifrons): *Bull. Fac. Fish. Nagasaki*, **2**, 8–25.

Padgett, W.J. and Tsokos, C.P. (1978), On Bayes estimation of reliability for mixtures of life distributions. *SIAM J. of Applied Maths*, **34**, no. 4, 692–703.

Parzen, E. (1962), On the estimation of a probability density function and the mode. *Annals of Math. Stats.*, **33**, 1065–76.

Paul, S.R. and Plackett, R.L. (1978), Inference sensitivity for Poisson mixtures. *Biometrika*, **65**, 591–602.

Pearson, K. (1894), Contribution to the mathematical theory of evolution. *Phil. Trans. A*, **185**, 71–110.

Pearson, K. (1914), A study of Trypanosome Strains. *Biometrika*, **10**, 85–143.

Pearson, K. (1915), On certain types of compound frequency distributions in which the components can be individually described by binomial series. *Biometrika*, **11**, 139–44.

Quandt, R.E. and Ramsey, J.B. (1978), Estimating mixtures of normal distributions and switching regressions. *J. Amer. Statist. Assoc.*, **73**, 730–38.

Rao, C.R. (1948), The utilization of multiple measurements in problems of biological classification. *J. Roy. Statist. Soc. Series B*, **10**, 159–203.

Rao, C.R. (1965), *Linear statistical inference and its applications*. John Wiley and Sons Inc., New York.

Rayment, P.R. (1972), The identification problem for a mixture of observations from two normal populations. *Technometrics*, **14**, 911–18.

Rider, P.R. (1961), The method of moments applied to a mixture of two exponential distributions. *Annals. of Math. Stats.*, **32**, 143–47.

Rider, P.R. (1962), Estimating the parameters of mixed Poisson, binomial and Weibull distributions by the method of moments. *Bull. Int. Statist. Institute*, **39**, 225–32.

Robbins, H. and Siegmund, D.O. (1974), Sequential tests involving two populations. *J. Amer. Statist. Assoc.*, **69**, 132–39.

Robertson, C.A. and Fryer, J.G. (1969), Some descriptive properties of normal mixtures. *Skand. Aktuarietidskr.*, **52**, 137–46.

Robertson, C.A. and Fryer, J.G. (1970), The bias and accuracy of moment estimators. *Biometrika*, **57**, 57–65.

Rosenblatt, M. (1956), Remarks on some nonparametric estimates of a density function. *Annals of Math. Stats.*, **27**, 832–37.

Ross, G.J.S. (1970), The efficient use of function minimization in nonlinear maximum likelihood estimation. *Applied Statistics*, **19**, 205–21.

Scheaffer, R.L. (1974), On direct versus inverse sampling from mixed populations. *Biometrics*, **30**, 187–98.

Scheffé, H. (1959), *The Analysis of Variance*. Wiley, New York.

Scott, A.J. and Symon, M.J. (1971), Clustering Methods based on Likelihood Ratio Criteria. *Biometrics*, **27**, 387–97.

Seeger, P. (1972), A note on partitioning a set of normal populations by their locations with respect to two controls. *Ann. Math. Stats.*, **43**, 2019–23.

Shah, B.K. (1963), A note on the method of moments applied to a mixture of two logistic populations. *J. of the M.S. University of Baroda (Science Number)*, **12**, 21–22.

Sichel, H.S. (1957), On the size distribution of airborne mine dust. *J.S.A. Inst. of Min. and Met.*, **58**, No. 5.

Silverman, B. (1978), Density ratios, empirical likelihood, and cot death. *Applied Statistics*, **27**, 26–33.

Smith, A.F.M. and Makov, U.E. (1978), A quasi-Bayes sequential procedure for mixtures. *J. Royal Statist. Soc., Series B*, **40**, 106–12.

Smith, M.R., Cohn-Sfetcu, S., and Buckmaster, H.A. (1976), Decomposition of multicomponent exponential decays by spectral analysis techniques. *Technometrics*, **18**, 467–82.

Steutel, F.W. (1967), Note on the infinite divisibility of exponential mixtures. *Ann. Math. Statist.*, **38**, 1303–05.

Steutel, F.W. (1970), Preservation of infinite divisibility under mixing and related topics, *Mathematical Centre Tracts*, **33**. Mathematisch Centrum, Amsterdam.

Subrahmaniam, K., Subrahmaniam, K. and Messeri, J.Y. (1975), On the robustness of some tests of significance in sampling from a compound normal population. *J. Amer. Statist. Assoc.*, **70**, 435–38.

Tallis, G.M. (1969), The identifiability of mixtures of distributions. *J. Appl. Prob.*, **6**, 389–98.

Tallis, G.M. and Light, R. (1968), The use of fractional moments for estimating the parameters of a mixed exponential distribution. *Technometrics*, **10**, 161–75.

Tan, W.Y. and Chang, W.C. (1972), Some comparisons of the method of moments and the method of maximum likelihood in estimating parameters of a mixture of two normal densities. *J. Amer. Statist. Assoc.*, **67**, 702–708.

Tarter, M. and Silvers, A. (1975), Implementation and application of bivariate Gaussian mixture decomposition. *J. Amer. Statist. Assoc.*, **70**, 47–55.

Teicher, H. (1960), On the mixture of distributions. *Annals of Math. Stats.*, **31**, 55–73.

Teicher, H. (1961), Identifiability of mixtures. *Annals of Math. Stats.*, **32**, 244–48.

Teicher, H. (1963), Identifiability of finite mixtures. *Ann. Math. Stats.*, **34**, 1265–69.

Thomas, E.A.C. (1966), Mathematical models for the clustered firing of single cortical neurones. *Brit. J. Math. Stat. Psychol.*, **19**, 151–62.

Thomas, E.A.C. (1969), Distribution free tests for mixed probability distributions. *Biometrika*, **56**, 475–84.

Titterington, D.M. (1976), Updating a diagnostic system using unconfirmed cases, *Applied Statistics*, **25**, 238–47.

Tretter, M.J. and Walster, G.W. (1975), Central and noncentral distributions of Wilks' statistic in MANOVA as mixtures of incomplete Beta functions. *Ann. Stats.*, **3**, 467–72.

Wegman, E.J. (1972), Nonparametric probability density estimation: I. A summary of available methods. *Technometrics*, **14**, 533–46.

Weibull, M. (1950), The distribution of the t and z variables in the case of stratified samples with individuals taken from normal parent populations with varying means. *Skandinavisk Aktuarietidskrift*, **33**, 137–67.

Wertz, W. (1978), *Statistical density estimation: A survey*, Gottingen, Vandenhoek, and Ruprecht, Monographs in Applied Statistics and Econometrics No. 13.

Whittaker, J. (1973), The Bhattacharyya matrix for the mixture of two distributions. *Biometrika*, **60**, 201–02.

Wilks, S.S. (1938), The large sample distribution of the likelihood ratio for testing composite hypotheses. *Ann. Math. Stat.*, **9**, 60–62.

Wolfe, J.H. (1965), A computer program for the maximum likelihood analysis of types. *Technical Bulletin*, 65–15, US Naval Personnel Research Activity, San Diego.

Wolfe, J.H. (1967), NORMIX; computational methods for estimating the parameters of multivariate normal mixtures of distributions. *Research Memorandum*, **SRM 68-2**, U.S. Naval Personnel Research Activity, San Diego.

Wolfe, J.H. (1969), Pattern clustering by multivariate mixture analysis. *Research Memorandum*, **SRM 69-17**, US Naval Personnel Research Activity, San Diego.

Wolfe, J.H. (1970), Pattern clustering by multivariate mixture analysis. *Multiv. Behav. Res.*, **5**, 329–50.

Wolfe, J.H. (1971), A Monte Carlo study of the sampling distribution of the likelihood ratio for mixtures of multinormal distributions. Naval Personnel and Training Research Laboratory. *Technical Bulletin*, **STB 72-2**, San Diego, California, USA.

Yakowitz, S.J. and Spragins, J.D. (1968), On the identifiability of finite mixtures. *Ann. Math. Stats.*, **39**, 209–14.

Young, T.Y. and Coraluppi, G., (1976), Stochastic estimation of a mixture of normal density functions using an information criterion. *IEEE transactions on Information Theory*, **IT-16**, No. 3 258–63.

Zacks, S. (1973), Sequential design for a fixed width interval estimation of the common mean of 2 normal distributions. 1. The case of one variance known. *J. Amer. Statist. Assoc.*, **68**, 422–27.

Zlokazov, V.B. (1978), UPEAK-Spectro-Oriented routine for mixture decomposition. *Computer Physics. Comm.*, **13**, 389–98.

Index